The author has written this
textbook and in it he emb
sical

INTERNATIONAL SERIES OF MONOGRAPHS IN
HEATING, VENTILATION AND REFRIGERATION

GENERAL EDITORS: N. S. BILLINGTON and E. OWER

VOLUME 1

FANS

FANS

BY

WILLIAM C. OSBORNE

PERGAMON PRESS

OXFORD · LONDON · EDINBURGH · NEW YORK
TORONTO · PARIS · BRAUNSCHWEIG

Pergamon Press Ltd., Headington Hill Hall, Oxford
4 & 5 Fitzroy Square, London W.1
Pergamon Press (Scotland) Ltd., 2 & 3 Teviot Place, Edinburgh 1
Pergamon Press Inc., 44–01 21st Street, Long Island City, New York 11101
Pergamon of Canada, Ltd., 6 Adelaide Street East, Toronto, Ontario
Pergamon Press S.A.R.L., 24 rue des Écoles, Paris 5ᵉ
Vieweg & Sohn GmbH, Burgplatz 1, Braunschweig

First edition 1966

Library of Congress Catalog Card No. 66–18408

PRINTED IN GREAT BRITAIN BY BELL AND BAIN LTD., GLASGOW
2622/66

CONTENTS

PREFACE

THIS book has been written as a textbook and has attempted to embrace the simple physical principles necessary for the proper application, control and design of fans. It does not set out to be an exhaustive work on the subject, nor is it a substitute for experience. It is intended for students preparing for the entry examinations of the appropriate professional engineering Institutions. In an attempt to make the work self-contained a certain amount of background matter, normally covered by standard textbooks, has been included. However, for the more interested reader, the appendix contains outline designs of two types of fan and suggestions for further reading.

Thanks are due to Mr. S. P. Hawes and Mr. E. Ower who kindly read the script and made many useful suggestions (whilst not being responsible for any opinions expressed herein) and to my wife who not only suffered silently during the writing of the book but also helped to read the proofs.

EDITORS' PREFACE

MODERN industrial civilization depends for its existence on man's control of his environment. Simple comfort requires that in most parts of the world buildings must be artificially heated or cooled during some part of the year. Rising standards of living have made people intolerant of the conditions of yesteryear in factories, offices and the home, and manufacturing processes themselves are requiring ever closer control of environment. Present-day air travel would be impossible without the air conditioning of aircraft.

Heating and air conditioning, then, has an essential contribution to make to the life of everyone—in the home, at work, while travelling or during recreation. These engineering services can account for between one-tenth and one-half of the total cost of a building, depending on their complexity and sophistication. They require expert design; and the number of skilled personnel is, almost everywhere, too small.

These, then, are the justifications for a series of textbooks dealing with the design of heating and air conditioning plant and equipment. The series is planned to include the following volumes:

Basic principles of heating and ventilating
Heating and cooling load calculation
Heating and hot water supply
Ventilation and air conditioning
Fuels and boilerhouse practice
Heat and mass transfer
Air flow
Fans
Dust and air cleaning
Refrigeration fundamentals
Refrigeration—commercial and domestic practice

The first five volumes cover the major aspects of the work of the heating and ventilating engineer. It is desirable that the reader should be familiar with the contents of the first two before proceeding to the others. The remaining five books cater for the student who may be more intimately concerned with some special topic, and each can be read independently of the others. The present treatment has allowed more detailed consideration of the subjects than has usually been possible in omnibus volumes.

The authors have taken as their starting point a basic training in general engineering such as may be acquired during the first years of apprenticeship. On this foundation, the specialist treatment is built and carried to a level approximating to that of a first degree. The graduate engineer or physicist who wishes to enter this field will also find the series useful, since he is introduced to new disciplines (for example, human physiology or climatology) and new applications of his fundamental knowledge, while some parts of his undergraduate course work are taken to much greater depth. Throughout the whole series, the practical applications are stressed.

The volumes do not pretend to cover the whole range of problems encountered in design, though a student who has mastered the basic principles embodied therein should be a competent engineer capable of handling a majority of the tasks he will meet. For the rest, practical experience backed by further study of more advanced texts will be essential.

LIST OF SYMBOLS

a vibration amplitude, dimension.
b barometric pressure, dimension, $cg/2W$.
c unit damping force, velocity of sound.
d diameter.
e direct strain.
f friction factor, frequency, stress.
g moisture content of air, gravitational acceleration.
i slope of a beam.
j $\sqrt{-1}$.
k loss factor, spring stiffness, wave number ω/c, constant.
l length, dimension.
m mass, dimension.
n speed of rotation, number of reflections.
p pressure.
q shear stress.
r radius.
s blade spacing $2\pi r/z$.
t time, temperature, thickness.
u tangential velocity, sound particle velocity.
v velocity.
w weight per unit volume, blade relative velocity, weight per unit length.
x distance.
y deflection.
z number of blades.
A psychrometric constant, area, general constant.
B general constant.
C general constant.
C_d coefficient of discharge.
C_D coefficient of drag.
C_L coefficient of lift.
D drag force.
E modulus of elasticity.

F	force.
G	modulus of rigidity.
H	head.
I	sound intensity, second moment of area.
J	polar second moment of area.
K	coefficient, constant.
L	level, mean free path length, lift force.
M	moment.
P	power, perimeter.
Q	volume flow rate, directivity factor.
R	gas constant, resistance.
S	area, scale factor.
T	absolute temperature, torque, transmissibility, reverberation time.
V	volume.
W	weight, sound power.
X	distance.
Z	section modulus.
α	angle of attack, absorption coefficient.
β	blade angle.
γ	ratio of specific heats of a gas.
δ	deflection.
η	efficiency.
θ	angle.
λ	power coefficient, wavelength.
μ	coefficient of viscosity.
ν	hub diameter/impeller diameter (hub ratio).
ξ	particle displacement.
ρ	density.
σ	Poisson's ratio.
τ	viscous shear stress.
ϕ	volume coefficient, phase angle.
ψ	pressure coefficient.
ω	angular velocity, angular frequency.
Σ	sum of.

Symbols may be defined in the text for a specific purpose.

CHAPTER 1

FLUID MECHANICS

1.1. BASIC IDEAS

Fan engineering is a specialized branch of Applied Mechanics. A large part of the design and operation of fans is basically fluid mechanics and it is of advantage to consider some of the basic concepts and their application, thus:

(a) conservation of matter, which may simply be stated for a fluid element as

(density × cross-sectional area × velocity) entering

= (density × cross-sectional area × velocity) leaving;

(b) conservation of energy, which in many practical problems involves accounting for the difference in energy level (usually in the form of head or pressure) between two points in the fluid flow;

(c) Newton's laws of motion, namely

(i) unless compelled by force to do otherwise, a body continues in a state of rest, or of uniform motion,

(ii) action of a force results in a rate of change of momentum in the direction of the line of action of that force, and proportional to the magnitude of the force, thus

$$F = k\frac{\mathrm{d}}{\mathrm{d}t}(mv)$$

In most problems the mass remains constant whilst the velocity changes, and

$$F = km\frac{\mathrm{d}}{\mathrm{d}t}(v)$$

For a change in velocity it can be said that force is proportional to the product of mass and acceleration. It is

1

convenient to use a system of units for which the value of the constant k is unity, and some such systems are

Force	Mass	Acceleration
poundal	pound	feet per second²
pound force	slug	feet per second²
dyne	gramme	centimetres per second²
newton	kilogramme	metres per second²

(iii) to every action there is an equal and opposite reaction. For example, any force acting on a body at rest is counteracted by a force of equal magnitude acting in the opposite direction;

(d) the gas laws, $pV = mRT$, where p is the gas pressure, V is the volume occupied by the gas, T is the absolute temperature of the gas ($T°$ Rankine $= 459·6 + t°$ Fahrenheit, or $T°$ Kelvin $= 273·3 + t°$ Celsius), and R is the gas constant in the appropriate units, $1545·4/M$ foot pounds force per pound mass degree Rankine, or $848/M$ kilogramme force metres per kilogramme mass degree Kelvin, M being the molecular weight of the gas.

1.2. AIR DENSITY

From the gas laws,

$$\text{density } \rho = \frac{m}{V} = \frac{p}{RT} \qquad (1.1)$$

and is seen to be a variable quantity according to conditions. In recording the performance of fans it is convenient to work in terms of some standard air density value and to make a correction for practical working conditions. Air is a mixture of gases having one variable constituent, namely water vapour. At any temperature air can absorb up to some maximum quantity of water vapour, when it is referred to as saturated. The amount of water vapour present in air at a given time is denoted by a quantity known as relative humidity, which is the ratio of the actual partial pressure of the water vapour present to the partial pressure at saturation expressed as a percentage.

Considering air as having two major constituents, dry air and water vapour; in a given sample both occupy the same volume at the same temperature, thus

$$\frac{m_a R_a}{p_a} = \frac{m_s R_s}{p_s}$$

where subscripts a and s refer to air and water vapour, and p_a and p_s are the partial pressures of the components. The proportion of water vapour to air,

$$g = \frac{m_s}{m_a} = \frac{p_s R_a}{p_a R_s} = \frac{p_s}{p_{at} - p_s} \cdot \frac{R_a}{R_s} \qquad (1.2)$$

The ratio R_a/R_s has the value of $0 \cdot 622$; p_{at} is atmospheric pressure.

The partial pressure of water vapour must be calculated from a knowledge of air condition. Often a psychrometer consisting of two thermometers, one of which has a wetted wick surrounding the bulb, is used to determine the moisture content of the air. Heat and mass transfer from the wetted bulb result in a lowering of the temperature, and from this temperature and that of the dry bulb, the partial pressure of the water vapour may be calculated from the empirical equation

$$p_s = p_s' - Ab(t - t') \qquad (1.3)$$

where $t = $ dry bulb temperature, $t' = $ wet bulb temperature, $b = $ barometric pressure, $p_s' = $ saturation vapour pressure at temperature t' (which may be read from hygrometric tables) and $A = $ an empirical constant having the values given below;

t and t'	A (t and t' above 32°F)
°F	$3 \cdot 7 \times 10^{-4}$ (sling psychrometer)
°F	$4 \cdot 44 \times 10^{-4}$ (screen psychrometer)
°C	$6 \cdot 65 \times 10^{-4}$ (sling psychrometer)
°C	$8 \cdot 0 \times 10^{-4}$ (screen psychrometer)

The difference in values between the two psychrometers is largely due to the fact that there is an air velocity of about 10 ft/s (3 m/s) over the wet bulb of the sling psychrometer.

The volume of unit mass of dry air (and also of the moisture g associated with it) will be

$$V_s = \frac{RT}{p_{at} - p_s} \tag{1.4}$$

Since the total mass will be $1+g$ units, the density of the mixture will be

$$\frac{1+g}{V_s} \tag{1.5}$$

Fans are normally tested in laboratories within a range of ambient temperature of from 50°F (10°C) to 80°F (30°C) and at the top of this range the difference in density of dry air and air fully saturated with water vapour is about 1·5%. Since air normally contains some moisture, clearly little error will result if the moisture content is assumed to remain at some constant value. One fan test code[1] defines standard air as having a density of 0·075 lb/ft³ (1·20 kg/m³) corresponding to air having a dry bulb temperature of 68°F (20°C), a relative humidity of 62%, and a barometric pressure b of 30 in. of mercury (1016·2 millibars). It is assumed that this gas mixture obeys the gas laws and that

$$\rho \propto \frac{p_{at}}{T} \propto \frac{b}{T}$$

1.3. HEAD AND PRESSURE

Although often used in the same context, these terms have specific meanings. Pressure is force per unit area, whilst head is the height of a column of fluid above a specified datum. The relationship between them is a simple one, for if a column of fluid of weight per unit volume w has a height H and cross-sectional area A, the pressure on the base

$$p = \frac{wHA}{A} = wH \tag{1.6}$$

When working with liquids it is generally possible to measure the head of fluid concerned. With gas flow this is not possible and it is usual to measure the pressure exerted by the gas on a column of liquid. In the case of air flow the liquid is water of constant specified condition. For example, water at 62°F (16·5°C) has a weight per unit volume of 62·36 lb/ft^3, and 1 in. of water represents a pressure of $\frac{1}{12}$ of 62·36 = 5·20 lbf/ft^2. Similarly, 1 mm of water represents a pressure of 1 kgf/m^2 at 4°C.

1.4. VELOCITY HEAD AND PRESSURE

If unit weight of fluid is raised by a distance H, it will have acquired an amount of potential energy of H units. Thus head may be regarded as energy per unit weight of fluid. If a weight of fluid W is moving with velocity v, it will possess kinetic energy per unit weight

$$H_v = \frac{Wv^2}{2gW} = \frac{v^2}{2g} \tag{1.7}$$

The quantity $v^2/2g$ is known as velocity head.
Velocity pressure

$$p_v = wH_v = \frac{wv^2}{2g} = \tfrac{1}{2}\rho v^2 \tag{1.8}$$

Velocity head is independent of air density whereas velocity pressure is not. It is important to observe the correct units, e.g.

v	ρ	g	H_v	p_v
ft/s	lb/ft^3	ft/s^2	ft	poundal/ft^2
cm/s	gm/cm^3	cm/s^2	cm	dyne/cm^2
m/s	kg/m^3	m/s^2	m	newton/m^2

It is more customary to work in
(i) v feet/minute and p_v inches of water

$$p_v = \frac{\rho}{2 \times 32 \cdot 2 \times 5 \cdot 2}\left(\frac{v}{60}\right)^2 = \rho\left(\frac{v}{1097}\right)^2 \text{ inches of water} \tag{1.9}$$

For air of standard density of 0·075 lb/ft^3, this becomes

$$p_v = \left(\frac{v}{4000}\right)^2 \text{ inches of water} \tag{1.10}$$

B

(ii) v metres/second and p_v millimetres of water

$$p_v = \frac{\rho v^2}{2 \times 9 \cdot 81} = \rho \left(\frac{v}{4 \cdot 43}\right)^2 \qquad (1.11)$$

kilogramme force/metre² or millimetres of water
For air of density of 1·20 kg/m³, this becomes

$$p_v = \left(\frac{v}{4 \cdot 03}\right)^2 \text{ millimetres of water} \qquad (1.12)$$

1.5. BERNOULLI'S EQUATION

Figure 1.1 shows an element of a stream tube in a fluid flow, i.e. an elemental tube in which flow is entirely parallel to the boundaries. For simplicity, it is assumed to have constant cross-sectional area of δa (although it can be shown it is not essential to do so). The forces

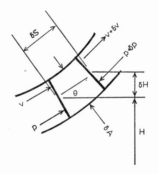

FIG. 1.1. Element of a stream tube.

on the element may be equated to the rate of change of momentum. In the direction of flow, the forces are,
due to change in pressure,

$$p\delta A - (p + \delta p)\delta A = -\delta p \delta A,$$

due to change in height above some datum,

$$-w\delta s \sin \theta \delta A = -w\delta H \delta A$$

Rate of change of momentum in direction of flow

$$= \frac{w}{g}\delta A v(v + \delta v) - \frac{w}{g}\delta A v^2 = \frac{w}{g}\delta A v \delta v$$

thus

$$-\delta p \delta A - w \delta H \delta A = \frac{w}{g} \delta A v \delta v$$

and rearranging,

$$v \delta v + \frac{\delta p g}{w} + g \delta H = 0$$

which in the limit becomes

$$v \mathrm{d}v + \frac{\mathrm{d}p}{\rho} + g \mathrm{d}H = 0$$

On integration, this gives

$$\frac{v^2}{2} + \int \frac{\mathrm{d}p}{\rho} + gH = \text{constant} \tag{1.13}$$

H is measured from any arbitrary datum, and any change of datum results in a change in H and an equal change in the constant of integration. In fan engineering, air is considered as incompressible and eqn. (1.13) reduces to

$$\frac{v^2}{2g} + \frac{p}{w} + H = \text{constant, known as total head} \tag{1.14}$$

Although strictly for flow along a stream tube of an ideal friction-less fluid, eqn. (1.14) is often used to relate conditions between two sections in a practical system of flow through a pipe. If the mean total head is measured at the two sections, it will be found that the value at the downstream section is less than that at the upstream section. This is due to resistance to flow between the sections and the difference in head is known as loss of total head. When making measurements, it is customary to use " gauge " pressures, i.e. pressures greater or less than atmospheric pressure. Considering two sections, subscript 1 referring to the upstream section and subscript 2 referring to the downstream section,

$$\frac{v_1^2}{2g} + \frac{p_{\text{at1}} + p_1}{w} + H_1 = \frac{v_2^2}{2g} + \frac{p_{\text{at2}} + p_2}{w} + H_2 + \Delta H \tag{1.15}$$

where ΔH is the loss of total head between the two sections. This may be rewritten

$$\frac{v_1^2}{2g} + \frac{p_1}{w} = \frac{v_2^2}{2g} + \frac{p_2}{w} + \Delta H + \left(H_2 - H_1 - \frac{p_{\text{at1}} - p_{\text{at2}}}{w} \right) \tag{1.16}$$

Now, if p_{at} represents the atmospheric pressure at a height H above some datum, and $p_{at}+\delta p_{at}$ at a height $H+\delta H$ above the same datum, and a column of air of cross-section A is considered,

$$p_{at}A-(p_{at}+\delta p_{at})A = wA(H+\delta H)-wAH$$

from which

$$-\delta p_{at} = w\delta H \tag{1.17}$$

If w remains constant, then eqn. (1.17) may be rewritten

$$\frac{p_{at1}-p_{at2}}{w} = H_2-H_1$$

and inserting this in eqn. (1.16) gives

$$\frac{v_1^2}{2g}+\frac{p_1}{w} = \frac{v_2^2}{2g}+\frac{p_2}{w}+\Delta H \tag{1.18}$$

Multiplying throughout by w gives the equation in terms of pressure;

$$\tfrac{1}{2}\rho v_1^2+p_1 = \tfrac{1}{2}\rho v_2^2+p_2+\Delta p \tag{1.19}$$

or

$$p_{t1} = p_{t2}+\Delta p$$

In eqn. (1.19), p_1 and p_2 are known as the static pressures at the two sections and may be positive or negative according to whether the absolute pressure is greater or less than the ambient atmospheric pressure which, as stated above, is the arbitrary datum or zero to which static pressure is generally referred. The sum of static pressure and velocity pressure $(p+\tfrac{1}{2}\rho v^2)$ is known as the total pressure p_t. Although in most practical cases the air density (and hence w) remains substantially constant, this may not be so where the height between two parts of a system is considerable, or if there is a temperature gradient.

Equation (1.19) shows that the resistance of a system of piping, expressed as a pressure loss for a particular flow rate, is equal to the difference between the total pressures at the two ends of the system. In practice the use of this equation to calculate the resistance of a system is complicated by the fact that the velocity nearly always varies considerably between the centre and the pipe walls, although the static pressure, except near bends, is often sensibly constant across a section. In determining the pressure loss it is not correct to calculate the velocity pressure component of the total pressure from the expression $\frac{1}{2}\rho v_m^2$, where v_m is the mean velocity and is equal to Q/A, Q being the volume flow and A the cross-sectional area of the airway. Strictly, as shown in reference 2, and neglecting any variations in the static pressure p across the section, the mean velocity pressure must be calculated from the kinetic energy per unit time divided by the volume flow per unit time, that is, in a circular pipe:

$$p_{v(\text{mean})} = \int\limits_0^R \tfrac{1}{2}\rho v \times v^2 \times 2\pi r \, dr \div \int\limits_0^R v \times 2\pi r \, dr$$

$$= \tfrac{1}{2}\rho \int\limits_0^R v^3 r \, dr \div \int\limits_0^R vr \, dr$$

In most cases where eqn. (1.19) is used, the error due to the incorrect method of calculating $p_{v(\text{mean})}$ is allowed for by an experimentally determined loss factor or coefficient for the form of velocity distribution it is hoped will be encountered. It will be assumed here that $p_{v(\text{mean})}$ is based on the simple calculation in conjunction with this factor.

1.6. MEASUREMENT OF PRESSURE

From eqn. (1.19) it is seen that the pressure in a duct in which air flows has two components. Velocity pressure results from the fluid inertia and acts in the direction of flow, whereas static pressure acts equally in all directions. Pressures are usually measured by inserting a tube into the flow, the other end of which is connected to a pressure measuring device which often registers the difference between the pressure in the duct and the ambient atmospheric pressure. It is essential that the pressure measuring tubes disturb the flow as little as possible. By suitable design[2] and orientation of the ends of the tubes open to the air flow, observations of total pressure and of static pressure can be made separately, and the velocity pressure is then the difference between the two. Static pressure is often measured with a side tube flush with, and at right angles to, the pipe wall; but certain precautions[2] are necessary if the results are to be free from error. This method cannot be used if the static pressure varies appreciably across the pipe section; and since, as already stated, the velocity nearly always varies, point readings across a section are generally essential to measure the mean rate of flow.

Point measurements may be made with a high degree of accuracy by using a pitot-static tube (Fig. 1.2), which consists of two concentric tubes, the inner one of which is open to provide a facing tube, whilst the outer is blanked off at the end but has orifices in the side

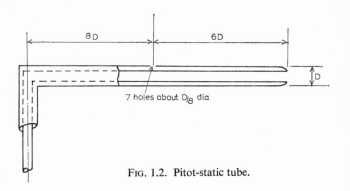

FIG. 1.2. Pitot-static tube.

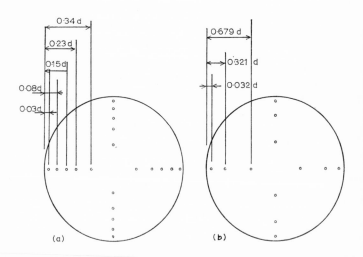

FIG. 1.3. Pitot tube traverses.
(a) 10 point tangential. (b) 6 point log–linear.

walls which are at right angles to the direction of flow. With appropriate connections, total, static or velocity pressure may be measured. Point readings are made on a traverse[1, 2, 3] and an average velocity calculated which, when multiplied by the cross-sectional area, gives the volume of air flowing in the duct. Figure 1.3 shows two forms of traverse in common use for ducts of circular cross-section, giving comparable accuracy. To obtain a similar degree of accuracy in ducts of rectangular cross-section, it is necessary to adopt traverses having many more points, it being important to explore the velocity distribution in the corners and around the walls of the duct.

At low air velocities, the velocity pressure is very small, and although sensitive micromanometers are available it is advantageous to adopt a device which will magnify the velocity pressure, even at the expense of some loss of total pressure. Bernoulli's equation may

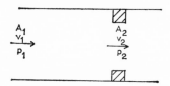

FIG. 1.4. Flow through a restriction in a pipe.

be used to show that a restriction in a duct will result in a magnified reading. From Fig. 1.4, by continuity, $A_1 v_1 = A_2 v_2$, from which $v_1 = A_2 v_2 / A_1$. Using Bernoulli's equation, and assuming no loss of total pressure,

$$p_1 + \tfrac{1}{2}\rho v_1^2 = p_2 + \tfrac{1}{2}\rho v_2^2$$

from which

$$p_1 - p_2 = \tfrac{1}{2}\rho (v_2^2 - v_1^2) = \tfrac{1}{2}\rho v_2^2 (1 - A_2^2 / A_1^2)$$

giving

$$v_2 = \sqrt{\left[\frac{2(p_1 - p_2)}{\rho (1 - A_2^2 / A_1^2)}\right]} \tag{1.20}$$

Volume flow,

$$Q = A_2 v_2 = A_2 \sqrt{\left[\frac{2(p_1 - p_2)}{\rho (1 - A_2^2 / A_1^2)}\right]} \tag{1.21}$$

The pressure difference, $\Delta p = p_1 - p_2$, is difficult to measure and two representative pressure tappings are often used to give a differential pressure and the volume flow is calculated with the aid of an experimentally determined discharge coefficient C_d, thus:

$$Q = C_d A_2 \sqrt{\left[\frac{2\Delta p}{\rho(1 - A_2^2/A_1^2)}\right]} \qquad (1.22)$$

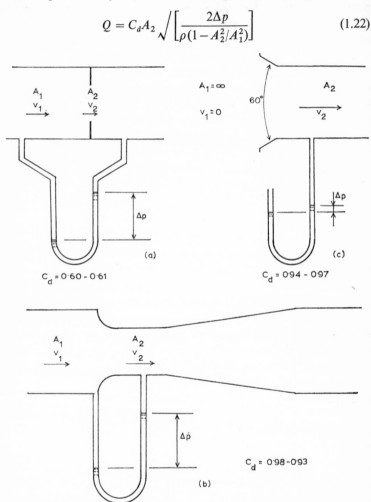

FIG. 1.5. Use of pipe restriction for flow measurement. (a) Orifice plate. (b) Venturi. (c) Inlet nozzle.

Some commonly used devices working on this principle are shown diagrammatically in Fig. 1.5. The loss of total pressure incurred is greater for the orifice plate of Fig. 1.5a than for the venturi of Fig.

1.5b which requires a greater length of airway. The entry nozzle of Fig. 1.5c may be regarded in a similar manner to a restriction if the area A_1 is infinitely large, when the area ratio, A_2/A_1 becomes zero. The resulting pressure differential is little greater than that given by a pitot-static tube but, when used with proper precautions, will give results to a similar degree of accuracy with a single reading instead of a rather tedious traverse.

1.7. FLUID FORCES ON SOLID BODIES

It may be observed that " layers " of fluid exert stresses on adjacent layers. For a fluid in motion the stress between adjacent layers is proportional to the local transverse velocity gradient. This phenomenon is known as *viscosity* and the constant of proportionality is known as the coefficient of viscosity μ. Thus

$$\mu = \tau \div dv/dy \tag{1.23}$$

where τ is the viscous shear stress in the fluid. At higher velocities, random motion of the fluid particles becomes evident. Consequently, inertia forces become predominant, the resulting energy being dissipated by viscous forces. On meeting a solid surface, the effect of the fluid inertia is to exert a force on the surface. By making reasonable assumptions, the dependence of these forces on certain parameters may be ascertained. For instance, it is to be expected that the effects will be dependent on fluid velocity v, on fluid characteristics as represented by coefficient of viscosity μ and fluid density ρ, and also the size of the surface (or body) as represented by a typical dimension l. Suppose the law relating fluid force F to these parameters may be written

$$F = kl^a v^b \rho^c \mu^d \tag{1.24}$$

where k is a numerical constant. The indices a, b, c and d may be determined by equating the dimensions of each side of the equation. Considering basic dimensions of length l, mass m and time t, the dimensions of each component will be

force; mass × acceleration is $m \times l \div t^2 = mlt^{-2}$,

length is l,

velocity; distance $\div$ time $= lt^{-1}$,

density; mass $\div$ volume $= m \div l^3 = ml^{-3}$,

coefficient of viscosity; (force ÷ area) ÷ (velocity ÷ distance)

$$= (mlt^{-2} \div l^2) \div (lt^{-1} \div l) = ml^{-1}t^{-1}.$$

Replacing eqn. (1.24) by its dimensions gives

$$mlt^{-2} = (l)^a(lt^{-1})^b(ml^{-3})^c(ml^{-1}t^{-1})^d$$

$$= l^{a+b-3c-d} \cdot t^{-b-d} \cdot m^{c+d}$$

Since the indices of m, l and t must be the same on each side of the equation, taking these in order:

t: $-2 = -b-d$, or $d = 2-b$;

m: $1 = c+d = c+2-b$, or $c = b-1$;

l: $1 = a+b-3c-d = a+b-3b+3-2+b = a-b+1$ or $a = b$.

Thus eqn. (1.24) may be written

$$F = kl^b v^b \rho^{b-1} \mu^{2-b} \tag{1.25}$$

and since it is found that, for most types of flow, inertia forces which vary as v^2 are the most significant, it is convenient to rewrite the equation in the form

$$F = kl^2 v^2 \rho \left(\frac{lv\rho}{\mu}\right)^{b-2} \tag{1.26}$$

The term $lv\rho/\mu$ is non-dimensional and is known as Reynolds number (Re). No single number for the index b has been observed experimentally. Equation (1.26) may thus be more conveniently written

$$F = kl^2 v^2 \rho \phi(Re) \tag{1.27}$$

where the function of Reynolds number $\phi(Re)$ is determined experimentally for each case of solid body or surface.

1.8. PRESSURE LOSS

In order to move a fluid through a system of ducts it is necessary to provide energy (a) to accelerate the fluid, i.e. kinetic energy, and (b) to overcome forces resisting the flow of fluid.

Pressure is energy per unit volume of fluid (for example, ft.lbf/ft^3 = lbf/ft^2) and so the total energy which has to be supplied by a fan or pump (and which is dissipated in the system) may be expressed in terms of volume flow Q and total pressure loss Δp.

So far as long straight pipes are concerned, the pressure loss may be deduced from eqn. (1.27) by considering Δp as the force per unit cross-sectional area and the pipe wall area (perimeter $P \times$ pipe length l) as the l^2, or surface area term. Thus

$$\Delta p = \frac{F}{A} = \frac{kl^2}{A} \rho v^2 \phi(Re) = f\frac{Pl}{A}\tfrac{1}{2}\rho v^2 \qquad (1.28)$$

where f is a " friction " factor with some dependence on Reynolds number, A is the pipe cross-sectional area, and $\tfrac{1}{2}\rho v^2$ is the velocity pressure corresponding to the mean flow velocity $v = Q/A$. The value of f also depends on the nature of the pipe surface, increasing as surface roughness increases, [4, 5, 6] and for galvanized sheet steel is of the order of 0·005.

Charts are available [7, 8] to facilitate the calculation of pressure loss in long straight ducts of circular cross-section. To use these for ducts of rectangular cross-section, it is convenient to find the equivalent diameter d_e for which the pressure loss per unit length is the same for a given flow volume and friction factor. For a rectangular duct of sides a and b

$$\frac{f\pi d_e l}{\pi d_e^2/4} \cdot \tfrac{1}{2}\rho \left(\frac{4Q}{\pi d_e^2}\right)^2 = f \cdot \frac{2(a+b)l}{a \times b} \cdot \tfrac{1}{2}\rho \left(\frac{Q}{a \times b}\right)^2$$

$$\pi d_e \left(\frac{4}{\pi d_e^2}\right)^3 = \frac{2(a+b)}{(a \times b)^3}$$

$$\frac{32}{\pi^2 d_e^5} = \frac{(a+b)}{(a \times b)^3}$$

$$d_e = \sqrt[5]{\frac{32(a \times b)^3}{\pi^2(a+b)}} = 1\cdot265 \sqrt[5]{\frac{(a \times b)^3}{a+b}} \qquad (1.29)$$

In many systems in which air is moved, it would appear that losses in straight ducts are unlikely to be the major source of pressure loss. At each change of direction, change of section area, obstruction due to equipment such as heaters and coolers and the like, pressure loss is evident. The fluid forces involved are due to inertia as for straight ducts, and it is usual, for each item of the system, to calculate the pressure loss by a simplified version of eqn. (1.27):

$$p = k \cdot p_v = k \cdot \tfrac{1}{2}\rho v^2 \qquad (1.30)$$

where k is an experimentally determined loss factor for the item concerned, and is regarded as having no strong dependence on (*Re*). For example:

(a) At entry to a duct, typical values based on the velocity pressure just inside the duct are: plain end, $k = 0.9$; flanged end, $k = 0.5$; $60°$ cone into end, $k = 0.2$. Where there are ornamental grilles covering the aperture, a loss factor appropriate to the form of obstruction must be used.

(b) At filters, heaters, coolers and the like, the value of k varies according to type and would generally be obtained from manufacturers' data.

(c) At bends there is pressure loss due to non-uniform velocity gradients in directions other than that of the mean flow and also to the necessity for the air to acquire momentum in a new direction. Some " flow separation " occurs resulting in eddy formation on the inside of the bend. For ducts of circular cross-section and of rectangular cross-section, some typical loss factors for different ratios of bend inside radius r to duct dimension (in the same plane) d are

r/d	k circular	k rectangular
0	0·8	1·0
0·25	0·4	0·4
0·5	0·25	0·2
1·0	0·16	0·13

The values for rectangular section ducts are valid for aspect ratios, b/d (b being the other dimension of the section), of up to about 3.

(d) At changes of cross-section, losses tend to be greater when there is an increase of section area in the direction of flow than where there is a reduction. For a sudden expansion it is possible, by making a simple assumption based on experimental evidence, to calculate the loss factor.

Referring to Fig. 1.6, it is found by experiment that p_0 is very nearly equal to p_1, and on this assumption the force exerted against the flow $= p_2 A_2 - [p_1 A_1 + p_0 (A_2 - A_1)] = (p_2 - p_1) A_2$. This must be equal to the rate of change of momentum in the same direction, $(wA_1 v_1^2/g) - (wA_2 v_2^2/g)$, and thus

$$(p_2-p_1)A_2 = \frac{wA_1v_1^2}{g} - \frac{wA_2v_2^2}{g} = \rho A_1 v_1^2 - \rho A_2 v_2^2 \qquad (1.31)$$

Now, by Bernoulli's equation, where Δp is the pressure loss,

$$p_1 + \tfrac{1}{2}\rho v_1^2 = p_2 + \tfrac{1}{2}\rho v_2^2 + \Delta p$$

$$\Delta p = p_1 - p_2 + \tfrac{1}{2}\rho v_1^2 - \tfrac{1}{2}\rho v_2^2 \qquad (1.32)$$

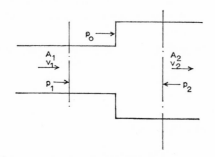

FIG. 1.6. Sudden expansion.

and combining eqns. (1.31) and (1.32)

$$\Delta p = \rho v_2^2 - \frac{\rho A_1}{A_2}v_1^2 + \tfrac{1}{2}\rho v_1^2 - \tfrac{1}{2}\rho v_2^2$$

$$= \tfrac{1}{2}\rho v_2^2 + \tfrac{1}{2}\rho v_1^2 - \rho \frac{A_1}{A_2}v_1^2$$

Now, $A_2 v_2 = A_1 v_1$, or

$$v_2 = \frac{A_1}{A_2}v_1$$

and thus

$$\Delta p = \tfrac{1}{2}\rho v_1^2\left(\frac{A_1^2}{A_2^2} - \frac{2A_1}{A_2} + 1\right)$$

$$= \tfrac{1}{2}\rho v_1^2\left(1 - \frac{A_1}{A_2}\right)^2 \quad [= \tfrac{1}{2}\rho(v_1-v_2)^2] \qquad (1.33)$$

From this equation it can be seen that the loss factor, based on the velocity pressure in the upstream duct section,

$$k = (1 - A_1/A_2)^2 \qquad (1.34)$$

For gradual expansions it is not easy to calculate the loss factor. It has been suggested[9] that this can be expressed with a fair degree of accuracy as $k = 0 \cdot 011 \theta^{1 \cdot 22}(1 - A_1/A_2)^2$, where θ is the included angle of the expansion expressed in degrees within the limits of $5°$ and $35°$.

The loss at contractions is rather less than at expansions. For a sudden contraction the loss factor based on the velocity pressure in the smaller section appears to vary from about $0 \cdot 25$ for an area ratio $A_1/A_2 = 2$ to $0 \cdot 37$ for an area ratio $A_1/A_2 = 4$. If the angle of convergence is less than about $60°$ (included), the loss factor reduces to about $0 \cdot 05$. For tapered changes of section without much change in area, and where the included angle does not exceed $60°$, a loss factor of $0 \cdot 15$ has been suggested.[7]

(e) At the pipe discharge it is customary to consider the velocity pressure as a loss since it must be supplied by the air moving device. If ornamental grilles cover the aperture, the additional loss must be accounted for, as in the case of inlet apertures.

(f) All systems in which air moves are closed circuits, the final link often being the ambient atmosphere. Apart from the discharge losses mentioned in (e) above, and the entry losses of paragraph (a), the atmospheric link is considered as a short-circuit with no further loss of pressure.

Some references to literature on the subject of pressure losses in air ducts will be found at the back of the book.

In single pipe or duct systems, the total pressure loss (that is, the necessary total pressure to be supplied by a fluid moving device) is found by adding the total pressure losses in individual sections, thus:

$$\Delta p = k_1 \cdot \tfrac{1}{2}\rho v_1^2 + k_2 \cdot \tfrac{1}{2}\rho v_2^2 + \ldots$$
$$= k_1 \cdot \tfrac{1}{2}\rho(Q/A_1)^2 + k_2 \cdot \tfrac{1}{2}\rho(Q/A_2)^2 + \ldots$$
$$= Q^2 \sum \frac{k\rho}{2A^2} \tag{1.35}$$

Once a system is designed and in operation, the summed parameters will be constant, and

$$\Delta p \propto Q^2 \tag{1.36}$$

A concept similar to that of electrical resistance is often useful, thus

$$\text{resistance } R = \frac{\Delta p}{Q^2} \tag{1.37}$$

Such a concept is used in mine ventilation, where it is customary to express the resistance of airways in atkinsons,[10] one atkinson being the resistance of a system for which the pressure loss Δp is 1 lbf/ft^2 for a flow of 1000 ft^3/s.

Following the electrical analogy, Fig. 1.7a shows resistances in series. The total pressure loss

$$\Delta p = \Delta p_1 + \Delta p_2 + \Delta p_3 = R_1 Q_1^2 + R_2 Q_2^2 + R_3 Q_3^2 = RQ^2$$

Now, for a series circuit, $Q_1 = Q_2 = Q_3 = Q$, and the equivalent resistance

$$R = R_1 + R_2 + R_3 \tag{1.38}$$

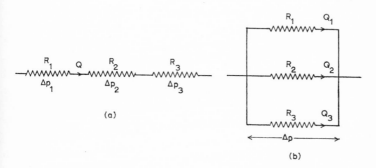

(a)

(b)

FIG. 1.7. (a) Flow resistances in series. (b) Flow resistances in parallel.

Figure 1.7b shows resistances in parallel, for which the total volume flow

$$Q = Q_1 + Q_2 + Q_3 = \sqrt{\left(\frac{\Delta p_1}{R_1}\right)} + \sqrt{\left(\frac{\Delta p_2}{R_2}\right)} + \sqrt{\left(\frac{\Delta p_3}{R_3}\right)} = \sqrt{\left(\frac{\Delta p}{R}\right)}$$

and, since for parallel circuits, there can be only one pressure loss across the common junctions, $\Delta p = \Delta p_1 = \Delta p_2 = \Delta p_3$.

$$\frac{1}{\sqrt{R}} = \frac{1}{\sqrt{R_1}} + \frac{1}{\sqrt{R_2}} + \frac{1}{\sqrt{R_3}} \tag{1.39}$$

from which the equivalent resistance R may be found.

More complex series/parallel circuits may be dealt with in this manner. Considering the network shown in Fig. 1.8, it is desirable to deal first with the parallel resistances R_2 and R_3.

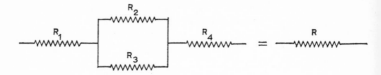

FIG. 1.8. Series/parallel network.

$$\frac{1}{\sqrt{R_{23}}} = \frac{1}{\sqrt{R_2}} + \frac{1}{\sqrt{R_3}} = \frac{\sqrt{R_3} + \sqrt{R_2}}{\sqrt{(R_2 R_3)}}$$

$$R_{23} = \frac{R_2 R_3}{R_2 + R_3 + 2\sqrt{(R_2 R_3)}}$$

$$R = R_1 + R_{23} + R_4 = R_1 + R_4 + \frac{R_2 R_3}{R_2 + R_3 + 2\sqrt{(R_2 R_3)}} \qquad (1.40)$$

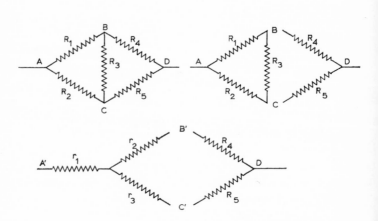

FIG. 1.9. Use of mesh–star transformation.

Another electrical analogy often useful is the mesh–star transformation[11] which can be applied to a problem such as is shown in

Fig. 1.9. The mesh ABC is transformed to its equivalent star connection $A'B'C'$, such that

$$R_{AB} = R_{A'B'}; \quad R_{BC} = R_{B'C'}; \quad R_{AC} = R_{A'C'}:$$

$$\frac{1}{\sqrt{R_{AB}}} = \frac{1}{\sqrt{R_1}} + \frac{1}{\sqrt{(R_2 + R_3)}} = \frac{\sqrt{(R_2 + R_3)} + \sqrt{R_1}}{\sqrt{[R_1(R_2 + R_3)]}}$$

$$R_{AB} = \frac{R_1(R_2 + R_3)}{R_1 + R_2 + R_3 + 2\sqrt{[R_1(R_2 + R_3)]}} = r_1 + r_2 \quad (1.41)$$

Similarly,

$$R_{AC} = \frac{R_2(R_1 + R_3)}{R_1 + R_2 + R_3 + 2\sqrt{[R_2(R_1 + R_3)]}} = r_1 + r_3 \quad (1.42)$$

and

$$R_{BC} = \frac{R_3(R_1 + R_2)}{R_1 + R_2 + R_3 + 2\sqrt{[R_3(R_1 + R_2)]}} = r_2 + r_3 \quad (1.43)$$

These equations may be solved for r_1, r_2 and r_3, in terms of R_1, R_2 and R_3 since

$$R_{AB} + R_{AC} - R_{BC} = 2r_2$$

$$R_{AB} + R_{BC} - R_{AC} = 2r_1$$

$$R_{BC} + R_{AC} - R_{AB} = 2r_3$$

In this way the mesh ABC is transformed into its equivalent star $A'B'C'$. The rest of the problem may be solved as for Fig. 1.8.

Other methods of expressing the resistance of air circuits include:

(a) Equivalent orifice, a hypothetical orifice in an infinite plate having a discharge coefficient of 0·65, for which the pressure differential is the same as the system total pressure loss at the same volume flow.

Putting $A_1 = \infty$, and $A_2/A_1 = 0$ in eqn. (1.22),

$$Q = 0.65 A_e \sqrt{\left(\frac{2\Delta p}{\rho}\right)} \quad (1.44)$$

The value of $\sqrt{(2\Delta p/\rho)}$ in practical units may be evaluated by using eqns. (1.9) to (1.12). The equivalent orifice is merely expressed as the area A_e in the appropriate units.

c

(b) Blast area, which is much the same as equivalent orifice but having a discharge coefficient of unity instead of 0·65.

1.9. EFFECT OF AIR DENSITY CHANGE ON PRESSURE LOSS

Equation (1.35) shows that, if the distribution of air density remains constant throughout a system,

$$\text{pressure loss } \Delta p \propto \rho Q^2 \tag{1.45}$$

Thus if only the ambient density changes (such as due to a change of barometric pressure), pressure loss at constant volume flow,

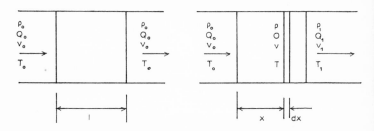

FIG. 1.10. Flow through a heat exchanger.

$\Delta p \propto \rho$. However, there may be local changes of air density at such components as heaters and coolers, and it is interesting to see the form of variation in pressure loss at these components as the density changes.

If the heater shown in Fig. 1.10 is assumed to be homogeneous and of constant cross-section, the pressure loss may reasonably be written, in accordance with eqn. (1.28), as

$$\Delta p_0 = kl \cdot \tfrac{1}{2}\rho_0 v_0^2 \tag{1.46}$$

where k is a constant representing pressure loss per unit length and velocity pressure, and subscript 0 represents some constant reference condition. If the density is not constant, as, for instance, when heat is applied, it is reasonable to write

$$\mathrm{d}(\Delta p) = k\mathrm{d}x \cdot \tfrac{1}{2}\rho v^2 \tag{1.47}$$

By continuity, $A\rho v = A\rho_0 v_0$, or $v = v_0\rho_0/\rho$

and

$$d(\Delta p) = k dx \cdot \tfrac{1}{2}\rho\left(\frac{\rho_0 v_0}{\rho}\right)^2$$

$$= k dx \frac{\rho_0}{\rho} \cdot \tfrac{1}{2}\rho_0 v_0^2$$

and substituting from eqn. (1.46) for $\Delta p_0/l = k \cdot \tfrac{1}{2}\rho_0 v_0^2$,

$$d(\Delta p) = \frac{\Delta p_0 \rho_0}{l\rho} dx \qquad (1.48)$$

The overall pressure loss may thus be found from a knowledge of the form of density variation through the heater by integrating eqn. (1.48) over the length of the unit. In practice, assumptions must be made as to the form of either the density or temperature distribution. The simplest form to consider is where the temperature varies uniformly along the length of a heater. In terms of absolute temperature

$$T = T_0 + \frac{x}{l}(T_1 - T_0) \qquad (1.49)$$

and, differentiating,

$$dT = \frac{dx}{l}(T_1 - T_0) \qquad (1.50)$$

Now, since $\rho \propto 1/T$ (eqn. (1.1)),

$$d(\Delta p) = \frac{\Delta p_0}{l} \cdot \frac{T}{T_0} \cdot dx$$

and, substituting the value of dx from eqn. (1.50),

$$d(\Delta p) = \Delta p_0 \frac{T \cdot dT}{T_0(T_1 - T_0)}$$

which on integration gives

$$\Delta p = \frac{\Delta p_0(T_1^2 - T_0^2)}{2T_0(T_1 - T_0)}$$

$$= \Delta p_0 \cdot \frac{T_1 + T_0}{2T_0} \qquad (1.51)$$

The pressure loss given by eqn. (1.51) is based on a volume flow of Q_0 entering the unit. Since the volume flow leaving the unit Q_1 will be, by continuity, $Q_0\rho_0/\rho_1$, the resistance appears to be different when considered from the downstream side of the heater (see the second example at the end of Chapter 3).

If a cooler is considered instead of a heater, eqn. (1.48) is still valid and eqn. (1.49), for a linear temperature distribution, may be written,

$$T = T_0 - \frac{x}{l}(T_0 - T_1) \tag{1.52}$$

which is seen to be the same as eqn. (1.49). Thus the result of eqn. (1.51) is still valid.

1.10. VORTEX FLOW

A vortex is a rotating mass of fluid, and is relevant to fluid flow in fans which set up such motion in the air passing through the impeller. The basic features of vortex motion may be considered by studying a fluid element in a vortex as indicated in Fig. 1.11.

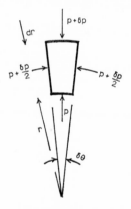

FIG. 1.11. Element of fluid in a vortex.

The element is subjected to forces due to pressure, p and $p + \delta p$ at radii of r and $r + \delta r$ respectively, and average pressures of $p + \frac{1}{2}\delta p$ on the radial faces. Considering the element to have unit thickness at right angles to the section shown, the net radial force inwards will be

$$(p + \delta p)(r + \delta r)\delta\theta - pr\delta\theta - 2(p + \tfrac{1}{2}\delta p)\delta r . \sin\tfrac{1}{2}\delta\theta$$

$$= pr\delta\theta + p\delta r\delta\theta + \delta pr\delta\theta + \delta p\delta r\delta\theta - pr\delta\theta - p\delta r\delta\theta - \tfrac{1}{2}\delta p\delta r\delta\theta$$

(since $\sin\frac{1}{2}\delta\theta$ becomes very nearly equal in value to $\frac{1}{2}\delta\theta$ as $\delta\theta$ approaches zero).

Neglecting third-order small quantities, net radial inward force on element $= r\delta p\delta\theta$.

Arising from centrifugal acceleration, $\omega^2 r\ (=u^2/r)$, resulting from the mass of fluid rotating at angular velocity ω, there will be a force on the element of

$$\rho\frac{u^2}{r}[\tfrac{1}{2}(r+\delta r)^2\delta\theta-\tfrac{1}{2}r^2\delta\theta]$$

$$=\rho\frac{u^2}{r}(\tfrac{1}{2}r^2\delta\theta+r\delta r\delta\theta+\tfrac{1}{2}\delta r^2\delta\theta-\tfrac{1}{2}r^2\delta\theta)$$

$$=\rho u^2\delta r\delta\theta \text{ (neglecting third-order small quantities)} \qquad (1.53)$$

For equilibrium of the fluid element

$$\rho u^2\delta r\delta\theta = r\delta p\delta\theta$$

and in the limit, as δr and δp become zero,

$$\frac{\mathrm{d}p}{\mathrm{d}r}=\frac{\rho u^2}{r} \qquad (1.54)$$

Three forms of vortex are of interest.

(a) A free vortex, where the total energy as expressed by Bernoulli's equation remains constant, that is

$$p+\tfrac{1}{2}\rho u^2 = \text{constant}$$

Differentiating this with respect to a radial direction

$$\frac{\mathrm{d}p}{\mathrm{d}r}+\rho u\frac{\mathrm{d}u}{\mathrm{d}r}=0 \qquad (1.55)$$

and substituting for $\mathrm{d}p/\mathrm{d}r$ from eqn. (1.54)

$$\rho\frac{u^2}{r}+\rho u\frac{\mathrm{d}u}{\mathrm{d}r}=0$$

$$\frac{\mathrm{d}r}{r}+\frac{\mathrm{d}u}{u}=0$$

On integration, this gives

$$\log_e r+\log_e u = \text{constant}$$

or

$$ur = \text{constant}$$

That is to say, tangential velocity varies inversely as radius.

(b) Forced vortex in which the fluid rotates at constant angular velocity. From eqn. (1.54)

$$\frac{dp}{dr} = \frac{\rho u^2}{r} = \rho \omega^2 r \qquad (1.56)$$

On integration, this becomes

$$p_2 - p_1 = \rho \frac{\omega^2}{2}(r_2^2 - r_1^2) = \tfrac{1}{2}\rho u_2^2 - \tfrac{1}{2}\rho u_1^2 \qquad (1.57)$$

In other words, the difference in static pressure across the vortex is equal to the difference in tangential velocity pressure.

(c) A vortex of the form $u^2 r =$ constant. It has been found by experiment that the vortex in a centrifugal dust separator follows this law very closely. From eqn. (1.54)

$$\frac{dp}{dr} = \rho \frac{u^2}{r} = \rho \frac{u^2 r}{r^2} = \rho \frac{\text{const}}{r^2} = \rho \frac{A}{r^2} \qquad (1.58)$$

$$dp = \rho A \frac{dr}{r^2} \quad \text{which, on integrating, becomes}$$

$$p_2 - p_1 = \rho A \cdot \left(\frac{1}{r_1} - \frac{1}{r_2}\right)$$

$$p_2 - p_1 = \rho u_1^2 - \rho u_2^2 \qquad (1.59)$$

This is useful in finding the pressure loss Δp in the vortex of a centrifugal separator for, by Bernoulli's equation,

$$p_1 + \tfrac{1}{2}\rho u_1^2 = p_2 + \tfrac{1}{2}\rho u_2^2 + \Delta p$$

$$\Delta p = p_1 - p_2 + \tfrac{1}{2}\rho u_1^2 - \tfrac{1}{2}\rho u_2^2$$

$$= \rho u_2^2 - \rho u_1^2 + \tfrac{1}{2}\rho u_1^2 - \tfrac{1}{2}\rho u_2^2$$

$$= \tfrac{1}{2}\rho u_2^2 - \tfrac{1}{2}\rho u_1^2 \qquad (1.60)$$

In eqn. (1.60), u_1 is the tangential velocity at the outer radius of the separator body (and is often very nearly the same as the linear velocity in the inlet duct) and u_2 is the tangential velocity at a radius of about half that of the exit duct, within which the law of the vortex is invalid.

1.11. FLOW OVER AEROFOILS

If a flat plate is inclined at an angle to a moving stream of air, there will be a net force exerted on the plate by the fluid of the form given in eqn. (1.27). This force, for convenience, may be split into components in the direction of flow, called drag, and at right angles to the direction of flow, called lift. Equation (1.27) may be extended to show the probable forms of these forces:

$$\text{Drag force } D = C_D . A . \tfrac{1}{2}\rho v^2 \tag{1.61}$$

$$\text{Lift force } L = C_L . A . \tfrac{1}{2}\rho v^2 \tag{1.62}$$

where A is some defined area (usually the surface area of the plate), and v is the mean flow velocity. C_D is known as the drag coefficient and C_L as the lift coefficient, both of these having some dependence on Reynolds number. If the angle of inclination of the plate to the flow, known as the angle of attack α, is altered, the forces, and consequently the coefficients C_D and C_L, may be expected to change.

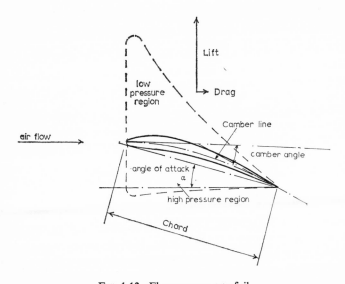

FIG. 1.12. Flow over an aerofoil.

The lift force may be looked upon as the useful component of the force whilst the drag is the energy loss component of the force. The aerofoil shape has been developed to give the highest ratio of lift to drag, primarily for aircraft which are sustained in flight when the lift force exactly balances the weight of the craft. The lift force is created by the shape of the upper surface which causes an increase in the air velocity locally and, by Bernoulli's equation, a reduction in static pressure. The local velocity at the under surface is little changed and a net upward force, or lift, results. Most of the lift force probably comes from the first 20% or so of the upper surface,

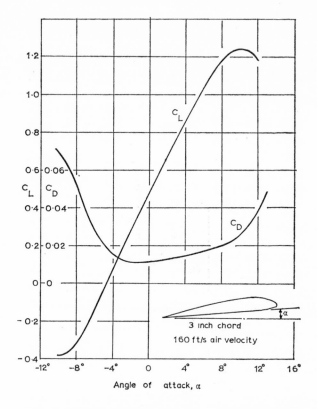

FIG. 1.13. Göttingen 436 aerofoil section performance details.

the remainder of this surface being shaped to give as little drag as possible by giving a very gradual return to normal flow conditions. Even so, if the angle of attack exceeds about 12–16° (depending on the type of section) severe flow separation occurs first at the trailing edge and then rapidly extending over almost the whole of the upper surface. The result of this is that the drag increases rapidly, whilst the lift reduces and may even fluctuate wildly. This phenomenon occurs quite suddenly and is known as " stalling ".

Many aerofoil shapes have been designed for various purposes, not all of which are particularly suitable for fan blade sections. Figure 1.13 shows data for a particular section (Göttingen 436) which has been used for axial flow fan design.

TABLE 1.1. Ordinates of Göttingen 436 Aerofoil. (Expressed as a percentage of the chord length.)

Station	Upper surface ordinate	Lower surface ordinate
0	2·66	2·66
1·25	4·53	1·21
2·5	5·54	0·79
5·0	7·00	0·37
7·5	8·11	0·15
10	8·98	0·05
15	10·16	0
20	10·82	0
30	11·08	0
40	10·55	0
50	9·60	0
60	8·28	0
70	6·60	0
80	4·70	0
90	2·64	0
95	1·45	0
100	0·25	0

1.12. EXAMPLES

1. Air at a temperature of 80°F and a barometric pressure of 29·25 in. of mercury flows at 75 ft/s. Find the velocity head and velocity pressure.

$$\text{Velocity head, } h_v = \frac{v^2}{2g} = \frac{75^2}{2 \times 32 \cdot 2} = 87 \cdot 3 \text{ ft.}$$

This represents the height to which a given weight of air must be raised in order to possess the same amount of potential energy as kinetic energy. This may be expressed in an equivalent height of water of weight per unit volume of 62·4 lb/ft³. The weight per unit volume of air is

$$0 \cdot 075 \times \frac{29 \cdot 25}{30} \times \frac{528}{460 + 80} = 0 \cdot 0715 \text{ lb/ft}^3$$

for the conditions stated (section 1.2). The equivalent head of water will be

$$\frac{0\cdot0715}{62\cdot4}\times87\cdot3 = 0\cdot10 \text{ ft}, \quad \text{or} \quad 1\cdot20 \text{ in.}$$

Velocity pressure is $p_v = \frac{1}{2}\rho v^2 = \frac{1}{2}\times0\cdot0715\times75^2$ (eqn. (1.8))

$$= 201 \text{ pdl/ft}^2$$
$$= 201/32\cdot2 = 6\cdot25 \text{ lbf/ft}^2.$$

This is equivalent to a head of water, since 1 in. of water exerts a pressure of $62\cdot4\times\frac{1}{12}= 5\cdot2$ lbf/ft^2, of $6\cdot25/5\cdot2 = 1\cdot20$ in.

2. A pitot-static tube traverse in a duct gives readings of velocity pressure at the centres of equal areas of 0·045, 0·110, 0·108 and 0·064 in. of water in flow where the air temperature is 65°F and the static pressure is 1·7 in. of water below the ambient barometric pressure of 30·3 in. of mercury. If the airway has an area of 4 ft^2, find the volume flow.

It is first necessary to find the average velocity from the point velocities. To find velocity from velocity pressure, $p_v = \frac{1}{2}\rho v^2$, thus, $v = \sqrt{[(2p_v)/\rho]}$, and for pressures in inches of water,

$$v = \sqrt{(2\times5\cdot2\times32\cdot2\times p_v/\rho)} \text{ ft/s}$$
$$= 60\sqrt{(2\times5\cdot2\times32\cdot2\times p_v/\rho)} \text{ ft/min}$$
$$= 1097\sqrt{(p_v/\rho)} \text{ ft/min.}$$

The density of the air within the airway will be

$$\rho = 0\cdot075\times\frac{30\cdot3-(1\cdot7/13\cdot6)}{30}\times\frac{528}{460+65} = 0\cdot076 \text{ lb/ft}^3.$$

Thus $v = (1097/\sqrt{0\cdot076})\sqrt{p_v} = 3980\sqrt{p_v}$ ft/min from which the point velocities are 843, 1320, 1310 and 1005 ft/min. The average velocity is $\frac{1}{4}\times4478 = 1119\cdot5$ ft/min.

Air volume flow = average velocity × area of airway

$$= 1119\cdot5\times4 = 4478 \text{ ft}^3/\text{min.}$$

3. A duct expands in cross-section from an area of 2 ft^2 to 3 ft^2. The mean total and static pressures immediately before the expansion are 1·4 in. and 1·8 in. of water respectively, and immediately after the expansion the mean total pressure is 1·5 in. of water, all pressures being below the ambient atmospheric pressure. Find the static pressure regain and the loss factor of the expander (referred to the velocity pressure in the smaller section).

Loss of pressure in the expander, $\Delta p = p_{t2}-p_{t1}$.

$$\Delta p = (p_{at}-1\cdot4)-(p_{at}-1\cdot5) = 0\cdot1 \text{ in. water.}$$

Applying Bernoulli's equation to upstream and downstream sections (eqn. (1.19)),

$$p_{t1} = p_{t2}+\Delta p,$$
$$p_1+\tfrac{1}{2}\rho v_1^2 = p_2+\tfrac{1}{2}\rho v_2^2+\Delta p.$$

Static regain, $p_2-p_1 = \tfrac{1}{2}\rho v_1^2 - \tfrac{1}{2}\rho v_2^2 - \Delta p.$

Now, $A_1v_1 = A_2v_2 = Q$, and thus $v_2 = v_1A_1/A_2$,

$$p_2-p_1 = \tfrac{1}{2}\rho v_1^2(1-A_1^2/A_2^2)-\Delta p,$$

$$p_{v1} = p_{t1}-p_1 = (p_{at}-1\cdot4)-(p_{at}-1\cdot8) = 0\cdot4 \text{ in. water,}$$

$$p_2-p_1 = 0\cdot4(1-2^2/3^2)-0\cdot1 = 0\cdot22-0\cdot1,$$

$$= 0\cdot12 \text{ in. water.}$$

Loss factor $= \Delta p/p_{v1} = 0\cdot1/0\cdot4 = 0\cdot25$.

4. A system consists of 100 ft of duct of cross-section 12 in. by 16 in., made from sheet steel whose friction factor is 0·005, two bends of the same cross-section each having a loss factor of 0·2, finally discharging through an expander of final dimensions 18 in. by 20 in. whose loss factor referred to the velocity pressure in the smaller section is 0·4. Find the total pressure loss for a volume flow of air of 2000 ft³/min having a density of 0·075 lb/ft³.

Component	Velocity $v = Q/A$ (ft/min)	Velocity press. $(v/4000)^2$ in.	Loss factor	Loss
Straight duct	1500	0·14	see below	0·245
Bends	1500	0·14	$2\times0\cdot2$	0·056
Expander	1500	0·14	0·4	0·056
Discharge	800	0·04	1	0·04
			Total	0·397

Loss factor for straight duct

$$= 0\cdot005\times\frac{2(16+12)144}{16\times12\times12}\times100 = 1\cdot75 \quad \text{(eqn. (1.28)).}$$

5. Assuming that the pressure loss in a centrifugal dust separator may be taken to be the sum of the entry pipe velocity pressure, the vortex loss, and a discharge loss corresponding to the highest velocity in the discharge pipe, show that the total pressure loss is $4\times r_1/r_e$ times the velocity pressure in the inlet pipe, where r_1 is the radius to the centre line of the inlet pipe, and r_e is the radius of the exit pipe. The vortex follows the law $u^2r = $ constant to a minimum radius of $r_e/2$, after which u reduces to zero.

Loss in inlet pipe $= \tfrac{1}{2}\rho v_1^2$.

Loss in vortex $= \tfrac{1}{2}\rho u_2^2-\tfrac{1}{2}\rho v_1^2$ (eqn. (1.60)).

Now, $u_2^2r_2 = u_1^2r_1$ from which $u_2^2 = u_1^2r_1/r_2 = 2u_1^2r_1/r_e$.

Thus loss in vortex $= \tfrac{1}{2}\rho v_1^2(2r_1/r_e-1)$ since $v_1 = u_1$.

Loss in exit pipe $= \tfrac{1}{2}\rho u_2^2 = \tfrac{1}{2}\rho v_1^2\times2r_1/r_e$.

Thus overall loss $= \tfrac{1}{2}\rho v_1^2+\tfrac{1}{2}\rho v_1^2(2r_1/r_e-1)+\tfrac{1}{2}\rho v_1^2\times2r_1/r_e$

$$= \tfrac{1}{2}\rho v_1^2\times4r_1/r_e.$$

CHAPTER 2

FANS

2.1. FAN PRESSURE, POWER AND EFFICIENCY

The purpose of a fan is to move air continuously against moderate pressures. Although under some circumstances a little compression occurs, it is customary to consider air in fans as being incompressible. For the few cases where error would result, correction factors may be applied.[1] Fan pressure is expressed in terms of pressure rise through the unit, rather than as a ratio of absolute pressures at outlet and inlet, as follows:

Fan total pressure p_t is the difference between the total pressures at the fan outlet and inlet;

Fan static pressure p_s is the fan total pressure minus the fan velocity pressure; and

Fan velocity pressure is the velocity pressure corresponding to the average velocity at the fan outlet (found by dividing the volume flow of air by the area of the fan discharge orifice).

It will be seen that the fan static pressure is not the rise in static pressure across the fan unit, but is defined in accordance with Bernoulli's equation. It is often regarded as the useful fan pressure.

The total pressure developed by a fan depends on the volume flow of air. When working with open inlet and outlet, the fan total pressure is equal to the fan velocity pressure, and the fan static pressure is zero. As the volume flow decreases, the fan total pressure (and consequently the fan static pressure) increases, reaching a maximum value at a point in the range of volume flow depending on the particular design of fan. The relationship between fan pressure and volume flow is determined by test. A graph of fan pressure against volume flow of air is known as the fan characteristic, the shape of which varies with type of fan. An example is shown in Fig. 2.1.

32

The work done in moving air against a constant pressure may be found by considering a piston of area A moving in a cylinder of air a distance l in time t against a constant pressure difference p. The work done will be pAl, while the power expended will be pAl/t. But Al/t is the volume of air Q moved in unit time, and thus

$$\text{air power } P = pQ \text{ in appropriate units} \qquad (2.1)$$

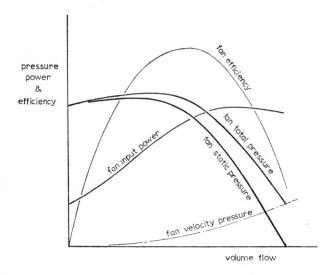

pressure
power
&
efficiency

fan efficiency

fan total pressure

fan input power

fan static pressure

fan velocity pressure

volume flow

FIG. 2.1. Fan characteristic curves.

If pressure p is measured in inches of water, this may be converted to lbf/ft^2 by multiplying by 5·2, and if the volume flow of air Q is in ft^3/min

$$\text{air power} = \frac{p \times 5 \cdot 2 \times Q}{33{,}000} \text{ h.p.} = \frac{p \times Q}{6360} \text{ h.p.} \qquad (2.2)$$

If pressure p is measured in mm water (kgf/m^2), and Q is in m^3/s,

$$\text{air power} = p \times Q \times 9 \cdot 81 \text{ watt} = \frac{p \times Q}{75} \text{ metric h.p.} \qquad (2.3)$$

The total power output of a fan is found by using fan total pressure in these equations, when the result is known as *air power* (*total*).[1] Occasionally, fan static pressure is used, the resulting power being known as *air power* (*static*).

Fan efficiency is the ratio of output power to mechanical input power and is usually expressed as a percentage:

$$\text{Fan total efficiency} \quad \eta = \frac{\text{air power (total)}}{\text{measured fan input power}} \times 100\% \quad (2.4)$$

$$\text{Fan static efficiency} \; \eta_s = \frac{\text{air power (static)}}{\text{measured fan input power}} \times 100\% \quad (2.5)$$

The measured fan input may be that absorbed by the impeller only, or may include the power absorbed by bearings and auxiliary drive components according to the design and use of the fan.

Because work is done on the air by a fan, there will be an increase in the internal energy, resulting in a rise in temperature. By equating the work done on the air in heat units to the heat quantity due to a rise in temperature, this latter may be found:

Fan power in heat units = mass × specific heat × temperature rise. Temperature rise

$$= \frac{\text{fan power in heat units}}{\text{air density} \times Q \times \text{specific heat}}$$

$$= \frac{p \times 5 \cdot 2 \times Q}{0 \cdot 075 \times Q \times 0 \cdot 24 \times 778} = 0 \cdot 37p \text{ deg F} \quad (2.6)$$

where p is the static pressure rise in the fan. Since fan inlet and outlet velocities are almost the same, p may be taken as fan total pressure with very little error. Alternatively, temperature rise

$$= \frac{p \times Q}{1 \cdot 2 \times Q \times 0 \cdot 24 \times 427} = 0 \cdot 0081p \text{ deg C} \quad (2.7)$$

where p is in millimetres of water. This does not take into account the energy loss in the fan which also results in a rise in air temperature. This point may be included in the above expressions by dividing by the fan total efficiency expressed as a fraction.

2.2. FAN LAWS

The performance of a fan in terms of pressure, volume flow and power absorbed depends on a number of factors, the most obvious of which are:

(a) the design and type of fan,
(b) the point of operation on the volume flow/pressure characteristic,
(c) the size of fan,
(d) the speed of rotation of the impeller,
(e) the condition of the air or gas passing through the fan.

It is customary for a manufacturer to make a range of fans of varying sizes to a single design, thus producing a series of geometrically similar fans (homologous series). It is convenient to be able to compute the performance of each fan from the minimum test data. The pressure/volume flow relationship is not generally capable of being expressed as a simple mathematical function. However, by considering any single point of operation on the characteristic curve (for example, the point at which the fan efficiency is a maximum), it is possible to derive some simple relationships, generally known as the fan laws.

The impeller diameter d may be used to indicate the fan size, and the peripheral speed of the impeller, u, to include the speed of rotation. The properties of the air (or gas) may be indicated by the coefficient of viscosity μ and the density ρ. As in section 1.7 it will be supposed that the laws relating volume flow Q and fan pressure p to these quantities may be written:

$$Q = k_q d^a u^b \rho^c \mu^d \tag{2.8}$$

$$p = k_p d^e u^f \rho^g \mu^h \tag{2.9}$$

where k_q and k_p are numerical constants.

Equating the dimensions of eqn. (2.8)

$$l^3 t^{-1} = (l)^a (lt^{-1})^b (ml^{-3})^c (ml^{-1}t^{-1})^d$$
$$= l^{a+b-3c-d} t^{-b-d} m^{c+d}$$

and taking the dimensions in convenient order and equating indices,

m: $0 = c+d$;

t: $-1 = -b-d$;

l: $3 = a+b-3c-d$;

from which $a = 2-d$; $b = 1-d$; $c = -d$; and eqn. (2.8) becomes

$$Q = k_q d^{2-d} u^{1-d} \rho^{-d} \mu^d = k_q d^2 u \left(\frac{du\rho}{\mu}\right)^{-d} \qquad (2.10)$$

Similarly, with eqn. (2.9)

$$ml^{-1}t^{-2} = (l)^e (lt^{-1})^f (ml^{-3})^g (ml^{-1}t^{-1})^h$$
$$= l^{e+f-3g-h} t^{-f-h} m^{g+h}$$

and, equating indices,

m: $1 = g+h$;

t: $-2 = -f-h$;

l: $-1 = e+f-3g-h$;

from which $e = -h$; $f = 2-h$; $g = 1-h$; and eqn. (2.9) becomes

$$p = k_p d^{-h} u^{2-h} \rho^{1-h} \mu^h = k_p u^2 \rho \left(\frac{du\rho}{\mu}\right)^{-h} \qquad (2.11)$$

In eqns. (2.10) and (2.11), the term $du\rho/\mu$ occurs, and it is seen to be of the same form as Reynolds number used in section 1.7. It may be taken as the Reynolds number of the fan (Re) based on the impeller diameter and peripheral velocity. If the speed of rotation of the impeller is n, then $u = \pi dn$, and eqns. (2.10) and (2.11) may be rewritten

$$Q = k_q d^3 n . f_1(Re) \qquad (2.12)$$

$$p = k_p d^2 n^2 \rho . f_2(Re) \qquad (2.13)$$

where $f_1(Re)$ and $f_2(Re)$ are variable factors based on Reynolds number. It is found that an increase in Reynolds number tends to result in an increase in fan performance and efficiency, although the effect is small. The fan laws are used mainly to predict the performance of fans in a homologous series from tests on a model of convenient size through which air from the test environment flows. One authority[1] permits calculation without correction for Reynolds number, provided that the Reynolds number under test conditions does not exceed that for the specified conditions by more than 40%.

Although not indicated by the dimensional analysis, other factors in practice tend to give increase of performance with increase in size. Relative roughness of the fan surfaces is generally less as size increases and blade profiles can be reproduced with greater accuracy. Also minor deviations from strict geometrical similarity inevitably occur in small fans where metal thickness and clearances tend to be relatively greater, resulting in degradation of performance. Compressibility effects cannot always be ignored for fan pressures of greater than 10 in. of water.

Although the effects mentioned in the previous paragraph are far from easy to predict, it is found that the fan laws may be simplified and still be capable of use with a good degree of accuracy over quite wide ranges of impeller diameter d, speed of rotation n, and air (or gas) density ρ, to

Volume flow
$$Q = k_q d^3 n \qquad (2.14)$$

Fan pressure
$$p = k_p d^2 n^2 \rho \qquad (2.15)$$

where fan pressure p may be either fan total, or fan static, pressure.

A third law may be derived since, from eqn. (2.1),

Fan power
$$P = pQ = k_P d^5 n^3 \rho \qquad (2.16)$$

The coefficients k_q, k_p and k_P will be constant for a range of geometrically similar fans, and for a particular point of operation on the pressure/volume characteristic.

2.3. PERFORMANCE COEFFICIENTS

Since the fan laws are valid for any particular point on the fan pressure/volume characteristic, similar laws will be valid for every other point of operation, the only difference being the numerical values of the coefficients. Thus a plot of k_p against k_q will have exactly the same form as the pressure/volume characteristic of each fan in the homologous series. It may be used, therefore, to represent the performance of any fan in the series, and may also be used to compare the performance of the series design with that of another series design. A system of performance coefficients based on this is[12]

Volume coefficient,
$$K_Q = \frac{Q}{d^3(n \text{ rev/min} \div 1000)} \qquad (2.17)$$

Pressure coefficient,

$$K_T = \frac{\text{fan total pressure } p_t}{d^2(n \text{ rev/min} \div 1000)^2} \qquad (2.18)$$

where fan pressure is for standard air conditions.

This coefficient has the subscript T when referring to fan total pressure, and subscript S when referring to fan static pressure.

Fan power coefficient

$$K_P = \frac{\text{fan h.p.}}{d^5(n \text{ rev/min} \div 1000)^3} \qquad (2.19)$$

In these expressions, Q is in ft³/min, fan pressure in inches of water and d in feet, thus the coefficients are not strictly non-dimensional. The values of k_q, k_p and k_P from eqns. (2.14) to (2.16) could, of course, be used, but would have inconvenient numerical values in most cases.

Commonly used non-dimensional coefficients based more nearly on eqns. (2.10) and (2.11) are

Volume coefficient

$$\phi = \frac{\text{volume flow}}{(\pi d^2/4)u} \qquad (2.20)$$

Total pressure coefficient

$$\psi = \frac{\text{fan total pressure}}{\frac{1}{2}\rho u^2} \qquad (2.21)$$

Static pressure coefficient

$$\psi_{st} = \frac{\text{fan static pressure}}{\frac{1}{2}\rho u^2} \qquad (2.22)$$

Power coefficient

$$\lambda = \frac{\phi\psi}{\eta} \qquad (2.23)$$

where d is the impeller diameter, and u the impeller peripheral velocity $(\pi d n)$.

A criterion used rather less with fans, perhaps, than for pumps, is specific speed n_s which is the speed at which a fan (of unspecified diameter) would run to give unit volume flow and unit fan pressure.

This applies to all fans in a homologous series, and is derived from the fan laws by eliminating the diameter term, thus:

$$p \propto \rho n^2 d^2$$

from which

$$d \propto \frac{p^{\frac{1}{2}}}{n \rho^{\frac{1}{2}}}$$

$$Q \propto nd^3 \propto \frac{np^{\frac{3}{2}}}{n^3 \rho^{\frac{3}{2}}} \propto \frac{p^{\frac{3}{2}}}{n^2 \rho^{\frac{3}{2}}}$$

$$n \propto \frac{p^{\frac{3}{4}}}{Q^{\frac{1}{2}} \rho^{\frac{3}{4}}}$$

and substituting n_s when $Q = 1$ and $p = 1$, the density remaining the same

$$\frac{n_s}{n} = \frac{Q^{\frac{1}{2}}}{p^{\frac{3}{4}}} \qquad (2.24)$$

These coefficients are almost always based on fan performance under standard air conditions.

2.4. CENTRIFUGAL FANS

It is probable that the majority of all fans are of the centrifugal or radial flow type. Such a fan consists of an impeller running in a casing having a spirally shaped contour (Fig. 2.2). The air enters the impeller in an axial direction and is discharged at the periphery, the impeller rotation being towards the casing outlet. The amount of work done on the air, evident in the pressure development of the fan, depends primarily on the angle of the fan blades with respect to the direction of rotation at the periphery of the impeller. Three main forms of blade are common (Fig. 2.2).

(a) Backward bladed, in which the blade tips incline away from the direction of rotation, and the blade angle β_2 is said to be less than 90°.
(b) Radial bladed, where the blade tips (or even the whole blade in the case of paddle bladed fans) are radial, that is, $\beta_2 = 90°$.
(c) Forward curved, where the blade tips incline towards the direction of rotation, and β_2 is greater than 90°.

The pressure developed increases with increase of blade angle, and for an impeller of given proportions, the volume flow also tends to increase. Forward curved fans generally develop the highest pressures for a given impeller diameter and speed, and they have also developed as high volume flow fans. This latter is evident by the

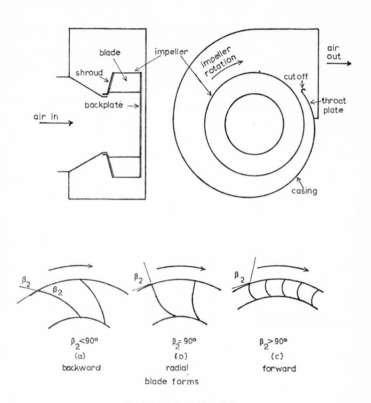

FIG. 2.2. Centrifugal fan.

small radial depth of the blades to allow as large an inlet diameter as possible to avoid " throttling " the higher volume flow entering the impeller. As a result of the shallower blades, the number of them has had to be increased in order to have the necessary influence on the air during its passage through the impeller. Even higher flow volumes may be obtained by having double inlet fans, these consisting effectively of two impellers back to back. Backward and radial bladed fans tend to have narrower impellers but fewer blades of greater radial depth—of the order of 6–16 in number in contrast to

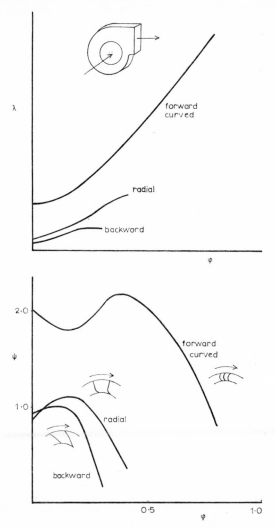

Fig. 2.3. Centrifugal fan characteristics.

the range of 30–60 of forward curved designs. Recently, backward bladed fans have been made with aerofoil section blades which has resulted in efficiencies of the order of 90%, compared with about 80%

for the best sheet metal designs. Efficiencies exceeding 75% are rare for radial bladed and forward curved fans. More complete details of some commercial fans are given in Table 5.1.

Most centrifugal fan impellers have shrouded blades, that is, annular plates are fitted at each end of the blades, giving mechanical strength to the impeller and reducing leakage between blades and casing. The paddle bladed fan impeller is unshrouded and consists of flat radial blades of substantial construction fitted to the drive hub. Such fans are used mostly for moving air containing solids since they remain reasonably free from blockage and can withstand considerable wear before failure. Forward curved fans are commonly used for ventilation of commercial buildings in view of their compact size for a given duty. Backward bladed fans are likely to be used for ventilation of large installations (for example, tunnels, coal mines) where their improved efficiency is of advantage. Radial tipped fans are often encountered as draught fans on large boilers, although some aerofoil bladed backward inclined fans are in use in this field.

Typical pressure/volume and power/volume characteristics for each type of fan are shown in Fig. 2.3, non-dimensional volume ϕ, pressure ψ and power λ coefficients being used to provide comparisons.

A study of the shape of the power/volume flow characteristics for forward curved (and also radial bladed fans) in Fig. 2.4 shows that a relatively small increase in volume flow may cause a considerable increase in power required. This may overload the electric motor drive since it is usually selected for a particular duty rather than for maximum possible fan power. For these types of fan it is not generally economical to select an electric motor (or any other prime mover) to drive such a fan under all conditions of operation, and usually a margin of 25–30% is allowed above the power estimated for the desired operating condition. It is recommended that ammeters be fitted to electric motor control gear to give an indication of the running conditions in order to avoid the possibility of overload. With backward curved fans, however, the maximum power required is generally little in excess of the power absorbed by the fan when working at maximum efficiency. Consequently, a motor able to cope with all operating conditions may economically be selected. This type of power/volume flow characteristic is often referred to as " non-overloading ".

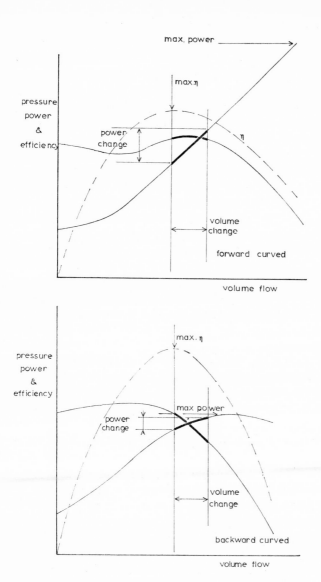

FIG. 2.4. Effect of a change in volume flow on centrifugal fan power.

2.5. AXIAL FLOW FANS

An axial flow fan may be described as a fan in which the flow of air is substantially parallel to the axis of the impeller. The tips of the impeller blades, which are commonly of aerofoil section, run with as fine a clearance as is practicable (consistent with cost of manufacture) in a cylindrical casing. In the simplest form of unit, air approaches the impeller in an axial direction and leaves with a rotational component due to work done by the impeller torque. Thus the absolute velocity of the leaving air is higher than the axial velocity with the result that some of the total pressure developed by the impeller does not appear as useful fan total pressure. More advanced designs have guide vanes downstream of the impeller which remove the rotational component, thus slowing down the air and converting some of the excess velocity pressure to more useful static pressure. Another way to achieve the maximum amount of useful pressure is to have pre-rotational vanes upstream of the impeller. These rotate the air in a direction opposite to that of the impeller rotation and with careful design the air will leave the fan in an axial direction. A third way to achieve the maximum useful pressure is to dispense with guide vanes and to have a second impeller downstream of the first, but rotating in the opposite direction. This second impeller then acts in a manner similar to that of an upstream guide vane unit, taking its pre-rotated air from the discharge of the first impeller. Such an assembly is known as a contra-rotating fan. True axial discharge from any of these units will be possible only for a single operating condition.

The pressure/volume flow characteristic of a fan with a high blade angle may exhibit a region of discontinuity corresponding to stalling conditions on the blade aerofoils. It is wise not to operate fans in this region, or at lower flow volumes. Upstream guide vane fans tend to show more marked stalling characteristics, although developing higher pressures, than downstream guide vane fans. The highest pressures may naturally be expected from contra-rotating fans, which often seem to show the least effects due to stall. Comparative characteristics for fans having impellers of similar proportions are shown in Fig. 2.5.

Although not developing as much pressure as centrifugal fans of the same impeller diameter and speed, axial flow fans may be used for applications involving the movement of air uncontaminated with

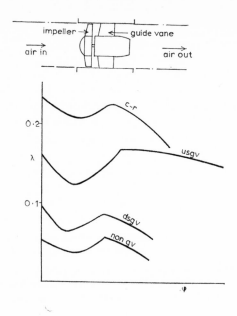

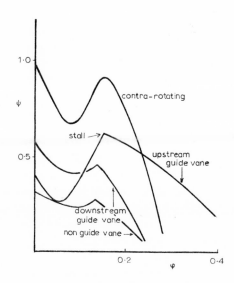

FIG. 2.5. Axial flow fan characteristics.

solids and at moderate temperatures (depending on the maximum permitted ambient temperature of the electric motor drive where this is in the air stream). Efficiencies of up to about 75% may be expected from non-guide vane fans, and up to about 87% for large downstream guide vane units.

A few mixed flow fans exist in which there is a radial component of flow (generally small) in addition to the axial component. The pressure development of these fans is comparable with the higher pressure axial flow fans.

2.6. PROPELLER FANS

Probably the simplest form of fan comprises a motor driving directly a sheet metal impeller, which runs with quite a large clearance in an orifice. Sometimes it may run open as a desk or ceiling fan merely to circulate air. Such a fan is a propeller fan. It should not be associated too closely with the axial flow fan since the air flow, rather than axial, is somewhat like that through a plain orifice. Units of moderate size (up to, say, 4 ft or 1·25 m) usually have 3–6

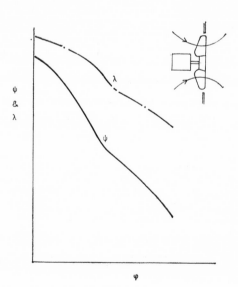

Fig. 2.6. Propeller fan characteristics.

blades of sheet metal mounted on a cast hub. Larger propeller fans
of up to 24 ft (7·5 m) or more in diameter are often really axial flow
fans in very short casings, since such a design is mechanically more
satisfactory.

It is not generally considered advisable to use propeller fans
against pressures greater than 0·5 in. (12·5 mm) of water, and the
usual field of application is to small unit equipments and simple
ventilation systems requiring little or no duct work. Large " propeller " fans are commonly used on cooling towers. Typical characteristics are shown in Fig. 2.6. Fan efficiencies range from about 60%
for sheet metal types to 75% for the larger units.

2.7. CROSS-FLOW FANS

The cross-flow, or tangential fan as it is sometimes called, has an
impeller with blades shaped somewhat like those of a forward curved
centrifugal fan impeller. However, both ends of the impeller are

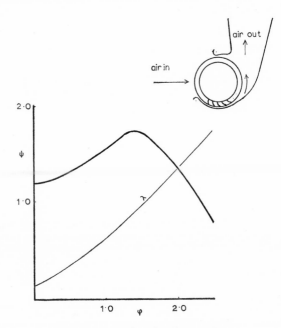

FIG. 2.7. Cross-flow fan characteristics.

sealed and it is fitted into a casing in which air enters at the periphery on one side, passes through the impeller, and leaves from the periphery at the other side. The flow is not diametral but along a curved path, the axes of inlet and outlet being roughly at right angles. Little development of large size units appears to have taken place, probably due to the fact that most of the pressure development is in the form of velocity pressure; also fan efficiency is low. However, in small sizes, they perform a useful function as is evident by their use in small domestic heaters. Volume flow is almost unlimited since the impeller may be made in any practicable length as throttling at the inlet (as in centrifugal fans) is not a problem. The characteristic of a small unit such as is used for domestic equipment is shown in Fig. 2.7. It should be pointed out that some authorities[13] report higher values of ψ than for forward curved fans.

2.8. EXAMPLES

1. A fan, whose outlet area is 3·5 ft², has a duty of 8000 ft³/min of air (of a density of 0·075 lb/ft³) at a fan total pressure of 2·50 in. of water with an input power of 4·10 h.p. Find the fan total and static efficiencies.

Average fan outlet velocity $= Q/A = 8000/3·5 = 2290$ ft/min.

Fan velocity pressure for the air condition stated (eqn. (1.10))

$$= (2290/4000)^2 = 0·33 \text{ in. water.}$$

Fan static pressure $= 2·50 - 0·33 = 2·17$ in. water.

$$\text{Fan total efficiency} = \frac{2·50 \times 5·2 \times 8000}{33,000 \times 4·10} \times 100\%$$

$$= \frac{2·50 \times 8000}{6350 \times 4·10} \times 100\% = 77\%.$$

$$\text{Fan static efficiency} = \frac{2·17 \times 8000}{6350 \times 4·10} \times 100\% = 67\%.$$

2. An axial flow fan, whose impeller has a diameter of 1·6 ft and runs at 1250 rev/min, has a duty of 2000 ft³/min at a fan static pressure of 0·70 in. of water for a fan total efficiency of 78% when the barometric pressure is 29·5 in. of mercury and the air temperature is 80°F. Find the volume flow and fan total pressure and input power at standard air conditions of 30 in. of mercury barometric pressure and 68°F for a geometrically similar fan whose impeller is 3·2 ft in diameter and runs at 1000 rev/min. Find also the specific speed and the values of ψ and ϕ.

Fan outlet area $= \dfrac{\pi}{4}(1·6)^2 = 2$ ft².

Average fan outlet velocity = 2000/2 = 1000 ft/min.

Fan velocity pressure = $\dfrac{29\cdot5}{30} \times \dfrac{528}{540} \times \left(\dfrac{1000}{4000}\right)^2 = 0\cdot06$ in. water.

Fan total pressure = $0\cdot70 + 0\cdot06 = 0\cdot76$ in. water.

Now, $Q \propto nd^3$, that is, $\dfrac{Q_2}{Q_1} = \dfrac{n_2 d_2^3}{n_2 d_1^3}$

$$Q_2 = Q_1 \frac{n_2 d_2^3}{n_1 d_1^3} = 2000 \times \frac{1000 \times 3\cdot2^3}{1250 \times 1\cdot6^3}$$

$$= 12{,}800 \text{ ft}^3/\text{min.}$$

Also $p \propto \rho n^2 d^2 \propto \dfrac{b}{T} n^2 d^2$, from which $p_2 = \dfrac{b_2 T_1 n_2^2 d_2^2}{b_1 T_2 n_1^2 d} \times p_1$

$$= 0\cdot76 \times \frac{30 \times 540}{29\cdot5 \times 528} \times \left(\frac{1000 \times 3\cdot2}{1250 \times 1\cdot6}\right)^2 = 2\cdot02 \text{ in. water.}$$

Fan input power = $\dfrac{2\cdot02 \times 12{,}800}{6350 \times 0\cdot78} = 5\cdot23$ h.p.

Fan static pressure of first fan under standard air conditions
$$= 0\cdot70 \times 30 \times 540/(29\cdot5 \times 528) = 0\cdot73 \text{ in. water.}$$

Specific speed based on this fan static pressure, $n_s = nQ^{\frac12}/p^{\frac34}$
$$= 1250 \times 2000^{\frac12}/0\cdot73^{\frac34} = 71{,}000 \text{ rev/min.}$$

Peripheral velocity pressure for larger fan $= (\pi \times 3\cdot2 \times 1000/4000)^2$

$$= \left(\frac{10{,}000}{4000}\right)^2 = 6\cdot30 \text{ in. water.}$$

$\psi = p/\frac12 \rho u^2 = 2\cdot02/6\cdot30 = 0\cdot321.$

$\phi = 4Q/\pi d^2 u = Q/Au = (4 \times 12{,}800)/(\pi \times 3\cdot2^2 \times 10{,}000) = 0\cdot16.$

3. Show, for a homologous series of fans running at constant impeller peripheral speed, that fan pressure remains constant, whilst volume flow and power both vary as the square of the impeller diameter.

Peripheral velocity = πdn = constant.
Fan pressure $\propto \rho d^2 n^2$ = constant for a given air density.
Volume flow $\propto d^3 n \propto dn \times d^2 \propto d^2.$
Fan power $\propto \rho d^5 n^3 \propto \rho d^3 n^3 \times d^2 u \propto d^2$ for constant air density.

CHAPTER 3

FAN OPERATION

3.1. FAN AND SYSTEM

The purpose of a fan is to move air at the required volume flow rate, and to supply the total pressure loss (including the discharge velocity pressure) in the system to which it is applied. Thus fan selection should be made on the basis of fan total pressure, which must include any pressure loss in the fan connections where these are of different size and shape from the duct connections.

It is usual for manufacturers to catalogue fan performances in terms of fan static pressure (under standard air conditions of 30 in. of mercury barometric pressure and a temperature of 68°F) and it is customary to select fans on this basis, thus:

Fan total pressure = system (total) pressure loss; or

Fan static pressure + fan velocity pressure = system pressure loss; or

Fan static pressure = system pressure loss − fan velocity pressure.

The assumption is often made that the fan velocity pressure is very nearly equal to the system discharge velocity pressure. In calculating system pressure loss, the discharge velocity pressure is ignored and the resulting total taken as being the required fan static pressure. Since, in many cases, the system pressure cannot be calculated precisely, any error may pass unnoticed; however, such an approximation should only be used with care. It is assumed that pressure loss has been calculated for standard air conditions. Where the working conditions differ from standard, an adjustment may be made by using eqn. (2.16). This may be substantial for fans passing hot gases (for example, induced draught fans on boiler plant).

Performance details supplied by fan manufacturers are based on tests carried out under ideal conditions of a particular test code.[1] When installed, the performance may be different from that predicted unless the fan connections are similar to those in the test code, usually long straight ducts at inlet and/or outlet. Bends, particularly at fan inlets, should be avoided, but if this is not possible, turning vanes should be fitted to ensure uniform velocity distribution over the duct cross-section. Where there is no duct connection to the fan inlet, it is advisable to fit an entry flare; this can be a short cone of 60° total angle having a length equal to about one-quarter of the fan inlet diameter. When used as part of unit equipment, the performance of a fan may be modified due to the proximity of other components and it is advisable to measure the overall performance of the assembly.

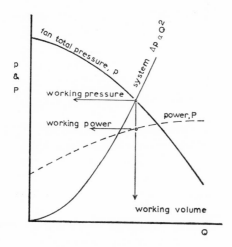

FIG. 3.1. Fan and system.

Fan manufacturers' catalogues cover very large ranges of volume flow of air and fan pressure in small increments to facilitate fan selection for single fixed duties. But it is often necessary to know the effects of changes in fan operation with, for example, modification to a system. The fan pressure/volume flow relationship is defined by the fan characteristic. The system pressure loss/volume flow relationship is defined by eqn. (1.36), that is, $\Delta p \propto Q^2$. If curves of these are plotted to the scales of pressure and volume flow on a single graph (Fig. 3.1), the intersection will give the point of operation. The fan power may be found by plotting the input power/volume flow relationship and reading off the power corresponding to the operating volume flow.

3.2. EFFECT OF CHANGE IN FAN SPEED

The fan laws, eqns. (2.14) and (2.15), show that at any operating point on the fan characteristic, *volume flow* $Q \propto$ *fan speed n*. Conversely, $n \propto Q$. *Fan pressure* $\propto n^2 \propto Q^2$. However, the *system pressure loss* also varies as Q^2 (to a sufficient degree of accuracy for most practical purposes). Thus the fan laws may be applied directly to

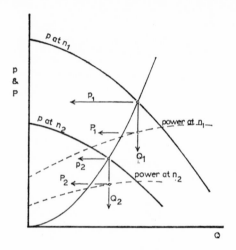

FIG. 3.2. Effect of speed change.

the point of operation of a fan on a system. From eqn. (2.16) it will be seen that *fan power* varies as n^3.

3.3. EFFECT OF CHANGE IN AIR DENSITY

If air passes through a system at uniform density which may change from one constant value to another, such as may occur with a change in ambient atmospheric conditions, both fan pressure (eqn. (2.15)) and system pressure loss (eqn. (1.35)) will vary proportionally to the air density, whilst fan volume flow will remain unchanged (eqn. (2.14)). Fan power will also be proportional to air density.

Where a change of air density occurs within the system, such as at a heater or cooler, the position of the fan may have some bearing on the result. The pressure loss in a heat exchanger with a linear

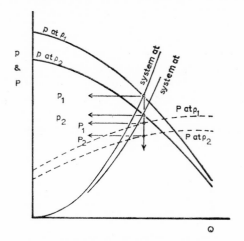

FIG. 3.3. Effect of change of air density.

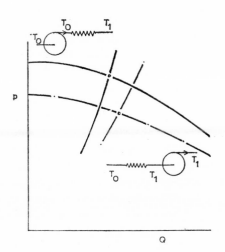

FIG. 3.4. Fan on a heat exchanger.

temperature gradient, based on the air volume flow entering, is given by eqn. (1.51) as

$$\Delta p = \Delta p_0 \times \frac{T_1 + T_0}{2T_0}$$

where Δp_0 is the pressure loss when leaving absolute temperature T_1 is the same as entering absolute temperature T_0. The volume flow of air leaving the heat exchanger will be $Q_0 \times T_1/T_0$. If the fan is placed on the upstream side of the heat exchanger, the pressure loss (against which the fan will have to supply total pressure) will be Δp for a volume flow of Q_0 at an absolute temperature T_0. If the fan is placed on the downstream side, the pressure loss will be Δp for a volume flow of air of $Q_0 T_1/T_0$ at an absolute temperature T_1. These represent two different operating conditions as shown in Fig. 3.4. The power absorbed by the fan will be that at the working volume flow and temperature. (See the example at the end of the chapter.)

3.4. FANS IN SERIES

When two or more fans are connected in series, the volume flow through each unit will be the same (since the air is regarded as being incompressible) whilst the overall total pressure will be the sum of the individual fan total pressures, less any losses in the interconnections. From this fact a combined characteristic for the assembly may be drawn, as shown in Fig. 3.5, by adding the fan total pressures at each volume flow for two fans. Here, in order to obtain a complete characteristic, it is assumed that each fan characteristic is known for flow volumes greater than those which are achieved by the fan when running with inlet and discharge unconnected to any system. Such information would rarely be available in practice since it would be necessary to assist the flow of air through the fan, a form of operation which has little point under normal working conditions. The free intake and discharge conditions of the assembly would be at point A or B according to whether fan 2 or fan 1 is the downstream unit providing the discharge velocity. If series operation of fans is chosen for an application, identical units should be used since it is unlikely that efficient operation would result otherwise. It is preferable, too, not to operate single impeller non-guide vane axial flow fans in series, unless these are widely spaced, since the rotational component of velocity in the air leaving the first impeller will modify operation of the second and succeeding fans.

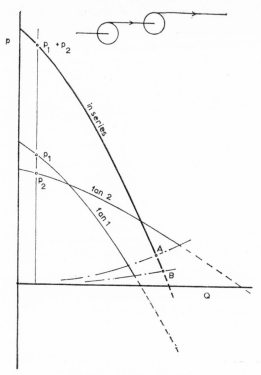

FIG. 3.5. Fans in series.

3.5. FANS IN PARALLEL

When two or more fans are connected in parallel there will be the same total pressure difference across each fan from the common connection point downstream to the common connection point upstream. The total volume flow will be the sum of the individual volumes flowing through each fan at the same effective fan total pressure, that is, allowing for loss of total pressure due to the individual fan connections. Based on this, a characteristic of the assembly may be drawn, as in Fig. 3.6 for two fans, by adding the volume flow of each fan at the same effective fan total pressure. In order to draw the combined characteristic in full it has been necessary to know the reverse flow characteristic of one of the fans (fan 2) with the impeller running in the normal direction. This would not normally be known since such operation is valueless. The free intake and discharge point of the assembly, point C, will depend on the size of the common discharge duct. As with series operation, identical fans would normally be used in parallel where this form of operation is desirable.

A convenient arrangement of fans in parallel is to mount each unit on the wall of a large chamber. If the cross-section A_2 of the chamber is large compared with the total fan discharge area A_1, the connection loss would be the same as for a sudden expansion and the loss factor, $k = (1 - A_1/A_2)^2$ (eqn. (1.34)), would tend to unity.

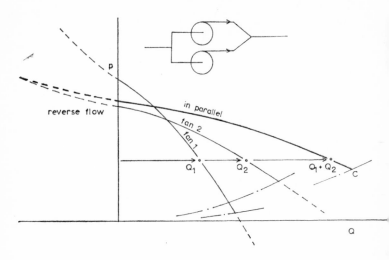

FIG. 3.6. Fans in parallel.

This means that there would be a pressure loss equal to the fan discharge velocity pressure. The effective total pressure across the common connections would then be fan total pressure minus fan velocity pressure, which is fan static pressure.

Not all fans have the simple characteristic as those used in Fig. 3.6. The characteristics of forward curved fans have points of contraflexure, with the result that as many as three different volume flow rates are possible at certain values of fan total pressure. As many as six volume flow rates are thus possible for two identical fans in parallel. Three of these values are merely twice the values for a single fan operating at the same total pressure ($2Q_x$, $2Q_y$ and $2Q_z$, for example in Fig. 3.7), and each fan works at the same relative point on its characteristic. The other three values ($Q_x + Q_y$, $Q_x + Q_z$, $Q_y + Q_z$) give rise to " appendages " on the combined characteristic

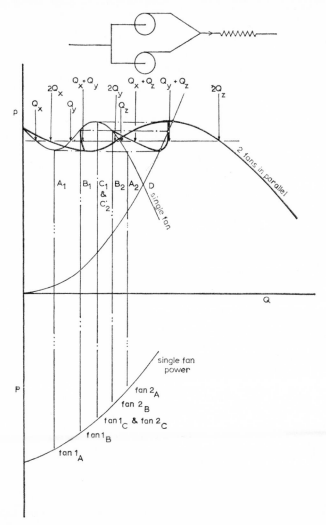

FIG. 3.7. Unsatisfactory operation of fans in parallel.

as shown in Fig. 3.7. If the fans so connected in parallel operate on a system having a characteristic such as that shown, there may be three points at which the volume flow and total pressure correspond

for both fan and system, such as *A*, *B* and *C*. It is impossible to fore-
cast the actual working condition and the fans may settle down at
any single point or oscillate between all three. Probably the most
serious effect of this situation is on the power consumption of the
motors driving the individual fans. The desired operating point
would be *C*, when both fans share the load equally as indicated by
points C_1 and C_2, and the motors would be selected accordingly. If
the point of operation is *B*, the load will be shared unequally, fan 1

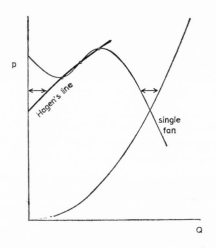

Fig. 3.8. Hagen's method of determining satisfactory operation of two fans
in parallel.

(point B_1) being underloaded whilst fan 2 (point B_2) is overloaded.
If the point of operation is *A*, the discrepancy is even greater, fan 2
(point A_2) being grossly overloaded. If the overload protection
devices in the motor control gear are correctly set, it is quite possible
that both motors will cut out immediately after starting up, the overloaded fan of the pair cutting out first, and the remaining fan then
operating at point *D*, where it will in turn be overloaded. In any
event it is wise to fit ammeters in the circuits to each fan motor to
indicate the degree of balance of power when fans are used in parallel.

A simple construction which may be used to determine whether or
not two similar fans will operate satisfactorily in parallel has been
suggested by Hagen.[14] This is shown in Fig. 3.8 and consists of

drawing the system characteristic and the characteristic of a single fan. The amount by which the system volume flow exceeds that given by the fan at any total pressure is plotted from the ordinate of pressure. If the resulting line cuts the characteristic of the single fan at more than one point, operation of two such fans on the system may be unsatisfactory.

Should trouble be encountered on existing installations, it may be possible to alleviate the condition by adding resistance, possibly in

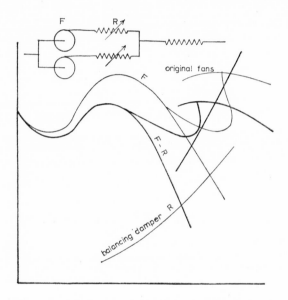

Fig. 3.9. Correction of unsatisfactory parallel operation.

the form of variable dampers, separately to each fan. The effect of this is shown in Fig. 3.9. The residual characteristic of fan and damper, $F-R$, is found by subtracting the pressure loss in the damper from the fan total pressure at each volume flow. The effect of this is to give a characteristic with a reduced " hump ". Thus careful adjustment of each damper will result in a combined characteristic (as observed on the ammeters) with balanced motor loads, and with little loss in volume flow or efficiency.

If fans in parallel are started one at a time, it is desirable to fit automatic backdraught shutters to prevent air recirculating through the fans not already running. This avoids reversed rotation of the impellers which could lead to heavy starting currents and mechanical shock on starting.

3.6. SIMPLE FAN AND SYSTEM NETWORKS

Many conventional air-conditioning systems are of the form shown basically by Fig. 3.10, comprising plant and main duct followed by a number of parallel distribution branches. To find the volume flow in each branch it is necessary to know first of all the total volume flow. It is most easily found by reducing the system to an equivalent single resistance. This may be solved using an equation similar to eqn. (1.40) or by the following method. Taking first the parallel

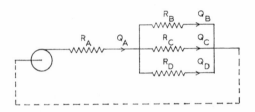

Fig. 3.10. Fan and series–parallel network.

branches, the pressure loss in each will have been calculated for the design flow volumes: for example, in branch B, Δp_B for a volume flow of Q_B; in branch C, Δp_C for a flow of Q_C; and in branch D, Δp_D for a flow of Q_D. It is unlikely that the pressure losses as calculated will be identical, but under operating conditions the pressure loss across each branch must be the same, whilst the total volume flow will be the sum of the individual branch flows. At this stage, the true volume flow, and hence the true pressure loss across the branches, is unknown. Consequently, some arbitrary pressure loss $\Delta p'$ is chosen, and the corresponding flows for this pressure loss calculated for each branch:

$$Q'_B = Q_B \times \sqrt{\Delta p'/\Delta p_B} \tag{3.1}$$

$$Q'_C = Q_C \times \sqrt{\Delta p'/\Delta p_C} \tag{3.2}$$

$$Q'_D = Q_D \times \sqrt{\Delta p'/\Delta p_D} \tag{3.3}$$

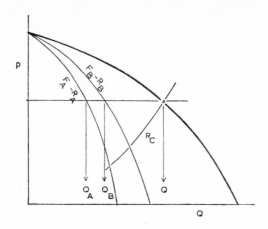

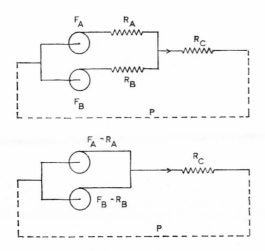

FIG. 3.11. Fans and systems in parallel.

(For convenience, $\Delta p'$ may be given the value of Δp_B, Δp_C or Δp_D.)
Thus the equivalent single resistance corresponding to the parallel
branches will have a pressure loss of $\Delta p'$ for a volume flow of
$Q'_{BCD} = Q'_B + Q'_C + Q'_D$. The system has now been reduced to two
series resistances, R_A and R_{BCD}. The volume flow through each of
these must be the same so, taking any arbitrary volume flow Q, the
pressure loss in R_A will be

$$\Delta p'_A = \Delta p_A \times (Q/Q_A)^2 \tag{3.4}$$

and the pressure loss in R_{BCD} will be

$$\Delta p'_{BCD} = \Delta p_{BCD} \times (Q/Q'_{BCD})^2 \tag{3.5}$$

The overall pressure loss in the equivalent single resistance will be
$\Delta p = \Delta p'_A + \Delta p'_{BCD}$ for a volume flow of Q.

Fan and system characteristics may now be plotted to find the
true volume flow through the main duct. Since $Q = \sqrt{(\Delta p/R)}$, and
Δp is the same for each parallel branch, the true volume flows in the
branches are in the same proportion as $Q'_B : Q'_C : Q'_D$. A numerical
example appears at the end of Chapter 1.

The system shown in Fig. 3.11 may conveniently be dealt with by
replacing fan F_A and resistance R_A with an " equivalent fan " having
the residual characteristic $F_A - R_A$ (as used in section 3.5). This is
found by subtracting the pressure loss in R_A from the fan total pressure for each value of volume flow. Similarly, the residual characteristic $F_B - R_B$ may be found. These two characteristics are combined in parallel (as for fans in parallel) and plotted to the same
ordinates as the characteristic of system R_C, the intersection giving
the working volume flow through R_C. The individual flows through
each fan may be read off the same diagram as shown.

Figure 3.12 shows in a somewhat different manner a system which
is basically the same as that of Fig. 3.11. R''_A and R'_A are in series
and may be combined into a single resistance R_A for the purpose of
finding the total flow through R_C. R''_B and R'_B may be treated
similarly. The pressure distribution around the network will be
different from the previous example, actual values being dependent
on knowing the absolute pressure at some defined point. For
example, point P is equivalent to the ambient atmospheric pressure
link in Fig. 3.11.

An interesting problem is that of supply and extract to a room
from which air may enter, or leave, by leakage (open doors, for

example). It is often desirable to know supply, extract and leakage flows, and also the pressure in the room. Considering first the supply fan and system alone, the volume flow and pressure in the room will be represented by some point on the residual characteristic, $F_A - R_A$. Similarly, considering only the extract system, the volume flow out of the room, and the pressure in the room will be represented by some point on the residual characteristic, $F_B - R_B$. If these two residual characteristics are plotted with respect to atmospheric pressure, that is, the supply system pressures positive and the extract pressures negative (Fig. 3.13), the intersection D will give the volume flow into the room when there is no leakage, and also the pressure in

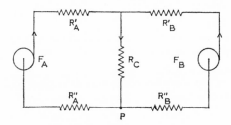

FIG. 3.12. Alternative equivalent network to that of Fig. 3.11.

the room relative to atmospheric pressure. It should be noted that it will have been necessary to continue the residual characteristic of one of the fans beyond the point of zero residual pressure. In the case in question, the more " powerful " fan and system supplies the pressure deficiency of the " weaker " fan and system. The maximum leakage from (or to) the room will be when the pressure in the room is the ambient atmospheric pressure, when the supply flow is represented by point F in Fig. 3.13, the extract flow by point E, and the leakage volume flow from the room by the difference between them. At pressures in the room between that at point D and atmospheric pressure, the supply and extract flows will be given by points at that pressure on lines DF and DE. From the point of view of leakage, the triangle DEF represents the room characteristic. This may be conveniently transferred to the normal pressure and volume flow ordinates by plotting the difference in volume flow between the lines DF and DE at each value of room pressure to give the characteristic $D'E'F'$. If the room leakage system characteristic R_C is now plotted

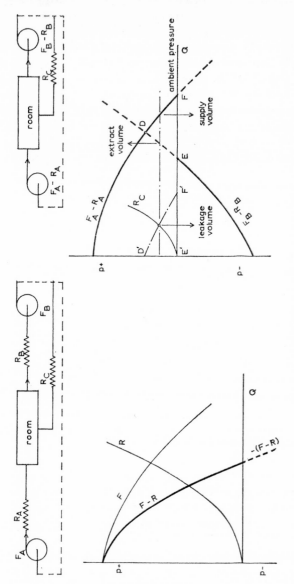

Fig. 3.13. Supply and exhaust of air to a room.

he intersection with line $D'E'F'$ gives the room pressure and leakage
olume flow. The supply and extract flows may be read off the
ppropriate curves at the room pressure. Although in Fig. 3.13 the
•oint D is at a pressure above atmospheric pressure, it could
•e below atmospheric with the appropriate fan and system
haracteristics.

The system shown in Fig. 3.14 may be seen to be the same problem
s that in Fig. 3.13 by considering the room to be at point Q and the

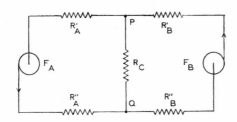

FIG. 3.14. Alternative equivalent network to Fig. 3.13.

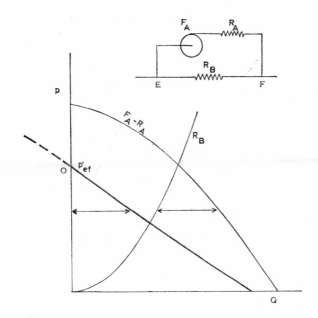

FIG. 3.15. Booster fan.

ambient atmosphere at point P. Having found the flows through each fan, the pressures at each point in the system may be found if the true pressure is known at any point.

Sometimes, for example in mine ventilation, it is necessary to have booster fans in a branch which may be in parallel with another branch as shown in Fig. 3.15. The residual characteristic of the boosted branch is indicated by $F_A - R_A$. The pressure difference from E to F will be the same for each branch and thus the surplus volume flow leaving the loop at point F for any pressure difference is found by subtracting the flow in the resistance from the flow in the boosted branch at that pressure, the resulting curve being the external characteristic of the loop. The dotted portion of the curve shown represents reverse flow in the direction F to E which could occur when the loop is part of a system supplied by an external fan which results in the pressure at F being greater than that at E by the amount p'_{ef}.

3.7. REVERSAL OF AIR FLOW

It is sometimes necessary or desirable to reverse the direction of air flow through a system. If axial flow fans are in use reversal of

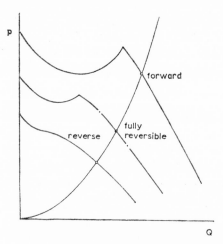

Fig. 3.16. Reverse flow characteristics of axial flow fans.

impeller rotation will give reversal of air flow. The fan pressure/volume flow characteristic will generally be less favourable under reverse flow conditions, and the efficiency will be less than for normal flow. Also, the system resistance may be somewhat different in reverse since, for example, reducing sections will become expansions. With non-guide vane fans of normal design the volume flow in reverse through a system will be of the order of 60–70% of the forward volume flow (Fig. 3.16). Guide vane fans would generally give less favourable reverse rotation characteristics. Some non-guide vane axial flow fans have been produced which give the same characteristics with either direction of rotation of the impeller. This is achieved by having some form of symmetrical blading. However, the performance of such a fan is generally inferior to that of a similar design for normal one-way operation.

Reversal of rotation of the impeller of a centrifugal fan does not result in reversal of direction of air flow, but merely a very much reduced flow in the normal direction. When using these fans, it is only possible to achieve flow reversal in a system by having an arrangement of ducts and variable baffles in which the flow path may be changed. Such an arrangement is shown diagrammatically in Fig. 3.17.

Fig. 3.17. Reversing circuit for a centrifugal fan.

3.8. REGULATION OF VOLUME FLOW

It has been shown in section 3.2 that the effect of a change in speed of rotation of a fan, when operating on a system for which the pressure loss is proportional to the square of the volume flow, may be predicted by the fan laws. Thus the operating point (which is preferably the point of maximum efficiency) remains unchanged. This is clearly the ideal method of volume regulation, although the arrangement used to vary the speed may introduce some power loss. Unfortunately, the most common, and probably least expensive, form of prime mover, the a.c. squirrel cage induction motor, is inherently a constant speed device. With special designs incorporating a high resistance rotor some speed regulation of up to about 50% of full

speed is possible by varying the supply voltage. However, the
increased rotor slip which gives the reduced speed also results in
increased rotor power losses which must be dissipated from the
motor in the form of heat. This method of speed regulation tends
to be limited in use to fractional horsepower motors with relatively
large frames for the rated output. Wound rotor induction motors
have been used with success, as have commutator motors (Schrage
motors, for example). If only two working conditions are sufficient,
a two-speed squirrel cage motor is less expensive. Where d.c. supplies
are available, speed control is more easily achieved, motors of the
series, or series compound, connected type being particularly suitable for a fan load torque which is proportional to the square of the
speed.

Speed variation may be achieved also by using a speed regulating
device between a constant speed motor and the fan. One of the
simplest of these, suitable for low power drives, is the variable speed
vee pulley, consisting of two halves sprung together on the motor
shaft. The motor is fitted on a sliding bedplate permitting the distance between the driving and driven pulleys to be altered, and consequently altering the pitch circle diameter of the driving pulley.
This gives a range of speed of about three to one. Alternatively, slip
devices such as magnetic and hydraulic couplings may be used,
although with inherent loss of efficiency. In these units the torque
on both halves of the coupling is the same. If the fan torque is T,
the motor power will be $2\pi n_0 T$, where n_0 is the constant motor speed.
Now fan power is proportional to fan speed to the third power, and
thus fan torque is proportional to n^2 and

$$\text{motor power} \propto n_0 T \propto n_0 n^2 \propto n^2 \tag{3.6}$$

If it were possible to have a coupling output speed equal to the input
speed n_0, the maximum fan (and motor) power would be $2\pi n_0 T_0$,
where T_0 is the fan torque at this speed; thus

$$\frac{\text{coupling power loss}}{\text{maximum fan power}} = \frac{2\pi(n_0 - n)T}{2\pi n_0 T_0} = \frac{(n_0 - n)n^2}{n_0^3}$$

since $T/T_0 = n^2/n_0^2$. This has a maximum value when $\mathrm{d}/\mathrm{d}n = 0$,
that is, when $2nn_0 = 3n^2$, or when $n = \frac{2}{3}.n_0$, of $(\frac{2}{3})^2 - (\frac{2}{3})^3 = 4/27$.
Thus a coupling of the slip type has a maximum power loss of 14·8%
of maximum possible power when the output speed is $\frac{2}{3}$ of the input
speed.

Variable speed devices tend to be expensive to install, consequently
ther methods of volume regulation are often adopted, the simplest
f which is throttling by means of an adjustable damper. The effect
f this is to increase the total system resistance, thus varying the
perating point on the fan characteristic. Although relatively low in
rst cost, it is not a very efficient method of regulation where a wide
ange of volume flow is required. Under these circumstances,
fficiency may be much improved by using a two-speed motor in
onjunction with the throttle as can be seen in Fig. 3.18. A more

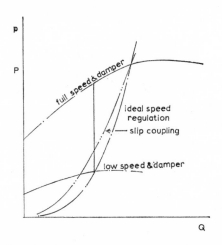

FIG. 3.18. Comparison of methods of volume flow regulation of a backward
curved fan.

fficient method, particularly suitable for use with centrifugal fans,
s to have inlet guide vanes to pre-rotate the air as it enters the fan
mpeller, and in the same direction of rotation. This has the effect of
reducing the amount of work done on the air by the impeller (see
section 5.1) and each guide vane setting results in a different fan
characteristic, illustrated by Fig. 3.19.

Fans have also been made with adjustable impellers, for example:
(a) Axial flow fans with variable or adjustable pitch blades.
(b) Centrifugal fans with retracting or variable angle blade tips.
(c) Centrifugal fans with variable width impellers.

F

These give volume regulation by variation of the pressure/volume flow (and power/volume flow) characteristics of a fan.

The choice of method of volume flow regulation will generally b a compromise between first cost and running cost, although ease o operation and installation may be significant factors in som instances, as may also be the possibility of automatic control.

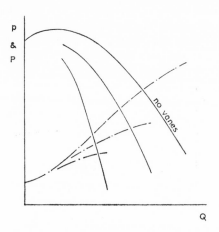

Fig. 3.19. Volume regulation of a radial tipped fan by variable inlet guide vanes.

3.9. EXAMPLES

These are based on fans having the following characteristics:

Volume flow ft³/min	Fan A		Fan B	
	Fan total pressure in. water	Fan input power h.p.	Fan static pressure in. water	Fan input power h.p.
0	3·00	0·83	3·00	2·40
2,000	3·02	1·43	2·76	2·70
4,000	2·92	2·23	2·68	3·45
6,000	2·36	2·90	2·97	4·60
8,000	1·10	2·90	2·85	6·00
10,000			2·28	7·60
12,000			1·48	9·25
14,000			0·54	11·00

1. Fan *A* supplies air to a system for which the total pressure loss is calculated to be 2·40 in. of water for a volume flow of air of 5000 ft³/min. Find the air volume flow and fan total efficiency.

The fan characteristic is plotted as shown in **Fig. 3.20.**

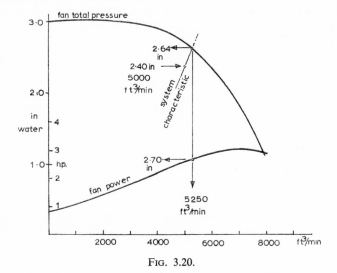

Fig. 3.20.

Also plotted is the system characteristic from the single point given. Since $\Delta p \propto Q^2$, $p_2 = p_1 (Q_2/Q_1)^2$.

For 4800 ft³/min, $\Delta p = 2·40 \times (4800/5000)^2 = 2·20$ in. water.
For 5200 ft³/min, $\Delta p = 2·40 \times (5200/5000)^2 = 2·60$ in. water.
The system characteristic intersects the fan characteristic at the point 5250 ft³/min, 2·64 in. water, and the power input at this volume flow is seen to be 2·70 h.p.

$$\text{Fan total efficiency} = \frac{2·64 \times 5250}{6350 \times 2·70} \times 100\% = 81\%.$$

2. Fan *A* is to be used to pass air through a heat exchanger for which the total pressure loss is measured as 2·20 in. of water for volume flow of air of 5000 ft³/min under laboratory conditions where the air flows at a constant temperature of 65°F. In operation air enters the heat exchanger at 60°F and leaves with a temperature of 200°F, being finally discharged to atmosphere through an area of 4 ft². Show where the fan should be placed to give the greater mass flow of air.

Pressure loss at a constant temperature of 60°F is

$$2·20 \times 525/520 = 2·22 \text{ in. water.}$$

Pressure loss with an inlet temperature of 60°F and an outlet temperature of 200°F is

$$2·23 \times \frac{520 + 660}{2 \times 520} = 2·53 \text{ in. water}$$

for a volume flow of air into the unit at 60°F of 5000 ft³/min and a volume flow out of the unit at 200°F of $5000 \times 660/520 = 6350$ ft³/min. The discharge velocity pressure under these conditions will be

$$\frac{528}{600} \times \left(\frac{6350}{4 \times 4000}\right)^2 = 0 \cdot 124 \text{ in. water.}$$

Thus the total pressure loss is $2 \cdot 53 + 0 \cdot 12 = 2 \cdot 65$ in. water.

The fan characteristics for 60°F and 200°F must be drawn. From the fan laws, volume flow at any operating point is independent of air density change, whilst fan pressure varies directly with air density. At 60°F the multiplying factor for fan pressure will be $528/520 = 1 \cdot 016$, and at 200°F, $528/660 = 0 \cdot 80$.

Volume flow (ft³/min)	Fan total pressure at 60°F in. water	Fan total pressure at 200°F in. water
0	3·05	2·40
2000	3·07	2·42
4000	2·97	2·34
6000	2·40	1·89
8000	1·02	0·88

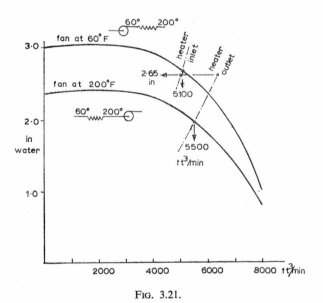

Fig. 3.21.

Case 1. Fan on the inlet side of the heater. The system pressure loss is 2·65 in. of water for a volume flow of 5000 ft³/min at 60°F. Plotting this characteristic with the fan characteristic at 60°F shows that the actual volume flow will be 5100 ft³/min. The air density at 60°F is

$$0.075 \times 528/520 = 0.0762 \text{ lb/ft}^3,$$

and the mass flow will be $5100 \times 0.0762 = 389$ lb/min. (See Fig. 3.21.)

Case 2. Fan on discharge (200°F) side of the heater. The system pressure loss is 2·65 in. of water for a volume flow of 6350 ft³/min at 200°F. Plotting this characteristic for 200°F shows that the volume flow will be 5500 ft³/min. The air density at 200°F is $0.075 \times 528/660 = 0.060$ lb/ft³, and the mass flow will be $5500 \times 0.060 = 330$ lb/min.

The mass flow of air is greater when the fan is on the low temperature side of the heat exchanger.

3. Fan *A* delivers air to a system, at a temperature of 68°F and 30 in. of mercury barometric pressure, at the rate of 5000 ft³/min. Show that, in order to obtain

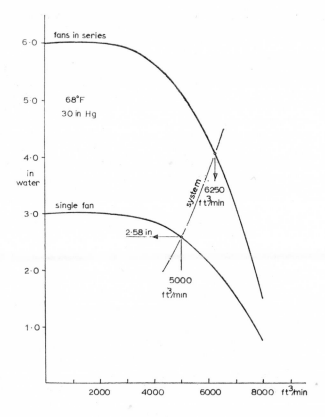

Fig. 3.22.

the same mass flow of air at 68°F where the barometric pressure is 24 in. of mercury, a second identical fan must be placed in series with the existing fan.

Assuming that the air density gradients in the system do not change, the change in fan pressure and change in system pressure loss at any volume flow will be in the same proportion, that is, in the ratio of the air densities, and the volume flow for the same fan and system arrangement will remain unaltered. The characteristics of a single fan, both fans in series, and the system pressure loss may all be drawn for the original temperature of 68°F and barometric pressure of 30 in. of mercury as shown in Fig. 3.22. The system characteristic is drawn through the point on the characteristic of the single fan where the volume flow is 5000 ft³/min, and it is seen that the volume flow from two fans in series will be 6250 ft³/min. At the barometric pressure of 30 in. of mercury the air density will be 0·075 lb/ft³ and the mass flow of air is

$$5000 \times 0·075 = 375 \text{ lb/min.}$$

At the barometric pressure of 24 in. of mercury the air density will be

$$0·075 \times 24/30 = 0·060 \text{ lb/ft}^3,$$

and the mass flow of air will be $6250 \times 0·060 = 375$ lb/min, the same as originally.

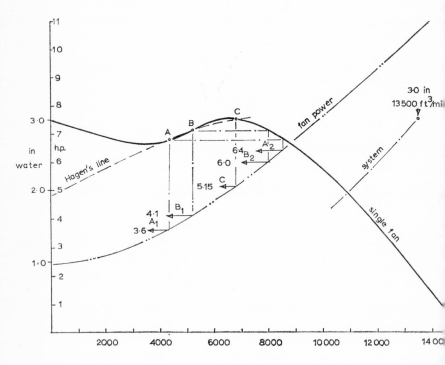

Fig. 3.23.

4. Two type *B* fans are to be used in parallel, fixed to a large chamber in which the air velocity is negligible, to deliver air to a system for which the pressure loss is 3·0 in. of water for an air flow of 13,500 ft³/min. Show whether or not they will operate satisfactorily.

Since the fans discharge into a large chamber in which the air velocity is negligible, the loss of pressure at this point is very nearly equal to the fan velocity pressure. Thus the effective fan total pressure is the fan total pressure minus the fan velocity pressure, that is, the fan static pressure. Plotting the characteristic based on this pressure for a single fan, and the system characteristic, Hagen's line may be constructed by plotting from the pressure ordinate a quantity equal to the volume required by the system minus the volume from a single fan at each appropriate pressure. It will be seen that Hagen's line cuts the fan characteristic at three points. This indicates that operation of two such fans on this system is likely to be unsatisfactory. The operating points for each fan under these conditions can be deduced and the power required by each fan read off the power curve as shown in Fig. 3.23.

5. One fan, type *A*, delivers air through a process system for which the pressure loss is 1·2 in. of water for an air flow of 6000 ft³/min, and a second fan of the same type delivers air through a second process system for which the pressure loss is 1·7 in. of water for a flow of 5000 ft³/min. A common system, for which the pressure loss is 1·2 in. of water for a flow of 11,000 ft³/min completes a closed circuit back to the fan inlet region. Find the flow through each process system, and the power input of each fan.

This arrangement is similar to that shown in Fig. 3.11. It is necessary to plot in parallel the residual characteristics of each fan and process system. (See Fig. 3.24.)

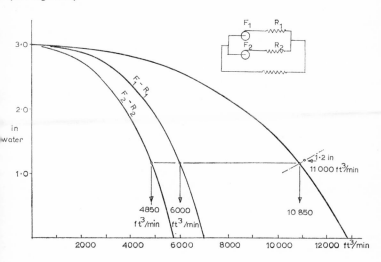

Fig. 3.24.

Volume flow (ft³/min)	Fan pressure F in. water	System pressure loss, R_1 in. water	$F - R_1$ in. water	System pressure loss, R_2 in. water	$F - R_2$ in. water
0	3·00	0	3·00	0	3·00
2000	3·02	0·13	2·89	0·27	2·75
4000	2·92	0·54	2·38	1·08	1·84
6000	2·36	1·20	1·16	2·45	(−0·09)
8000	1·10	2·13	(−1·03)	4·35	(−3·25)

Characteristics for $F - R_1$ and $F - R_2$ are plotted and the combined charac teristic in parallel derived by adding volumes at each pressure. The return system characteristic is plotted through the point 1·2 in., 11,000 ft³/min, and is found to intersect the parallel characteristic at a point 1·17 in., 10,850 ft³/min The individual system volumes are read off the respective residual characteristic curves for a pressure of 1·17 in. of water as 6000 ft³/min and 4850 ft³/min respectively.

6. A large process workshop is to be ventilated by a supply fan and system in which the pressure loss is 1·2 in. of water for a flow of 6000 ft³/min and an extract fan and system for which the pressure loss is 1·7 in. of water for a flow o 5000 ft³/min. Find the pressure in the workshop and the air volume flow. If there is a loss of air from the space by exfiltration through an aperture, where the pressure loss is 0·5 in. of water for a flow of 1000 ft³/min, find the pressure in the space, the supply volume flow, and the extract volume flow. The fans are type A.

Since the supply and extract systems in this question have the same resist ances as those in the last example, the calculations may be used. This problem is similar to that illustrated in Fig. 3.13. The residual characteristic for the supply fan and system, $F - R_1$, will generally tend to produce a pressure in the room above atmospheric pressure and is plotted as a positive pressure Similarly, the residual characteristic of the extract fan and system, $F - R_2$, will generally tend to produce a pressure in the room below atmospheric pressure and is plotted as a negative pressure. It is necessary in this case to plot the part of the residual characteristic representing a pressure deficiency (resulting from inadequate fan pressure to deliver higher volumes through the system and shown negative in the previous example). It is seen that the two charac teristics agree at the point 0·73 in., 6500 ft³/min, which will be the pressure in the workshop and the volume flow respectively.

The small triangle between the zero pressure line and the intersection of the residual characteristics may be regarded as the " room characteristic ". This is conveniently replotted from the pressure ordinate, together with the system characteristic through the point 0·5 in., 1000 ft³/min, and is seen, in Fig. 3.25 to result in an exfiltration flow of 750 ft³/min and a space pressure of 0·28 in of water. At this pressure it is seen that the supply air flow is 6950 ft³/min and that the extract flow is 6200 ft³/min.

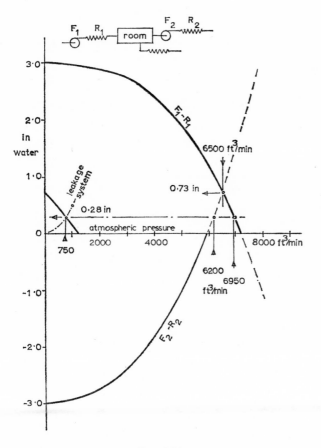

FIG. 3.25.

CHAPTER 4

VIBRATION AND NOISE IN
FAN SYSTEMS

4.1. SOURCES OF VIBRATION IN FANS

Probably the chief source of vibration in fans is due to rotational out-of-balance. In the case of fans which are short axially, the only likely form of unbalance to give trouble is *static unbalance*. This arises from lack of coincidence between the centre of gravity of the impeller and the axis of rotation, which gives rise to an unbalanced moment whose maximum value may be represented by $W_i.r = M$, W_i being the weight of the impeller and r the distance between impeller centre of gravity and the axis of rotation. When the impeller is rotating with angular velocity ω, there will be an outward radial force in the direction of radius r of

$$F = \frac{M}{g}\omega^2 \qquad (4.1)$$

Along any diametral line through the axis of rotation, the force will be

$$F \cos \omega t = \frac{M}{g}\omega^2 \cos \omega t \qquad (4.2)$$

where ωt is the angular displacement of the centre of gravity from the fixed line considered. The frequency of forced vibration f, that is, the number of cycles of force per unit time, will be the same as the number of revolutions in the same unit time, or

$$f = \frac{\omega}{2\pi} \qquad (4.3)$$

During the production of fans, static unbalance is eliminated as far as is practicable by mounting the impeller with its shaft horizontal in very free bearings (or on knife edges) and balancing out any tendency of the shaft to come to rest in any single attitude by adding weight to some convenient part of the impeller.

With impellers which are long axially, it is possible to achieve static balance by opposing moments which are not in the same radial plane. When the assembly rotates, an oscillating transverse couple is set up (Fig. 4.1) resulting in vibration from what is known as *dynamic out-of-balance*. This is eliminated as far as is practicable during manufacture, although methods of achieving dynamic balance are less simple than those for static balance. Essentially, two planes are chosen to which additional masses are added until axial vibration ceases, but at the same time preserving static balance. The frequency of vibration due to dynamic unbalance is again numerically the same as the speed of rotation.

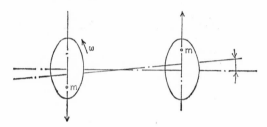

Fig. 4.1. Dynamic unbalance.

Torsional oscillations may be caused by aerodynamic forces in certain high speed fans, particularly when operating away from design conditions. Trouble from such sources in ventilating fans, which have relatively short, stiff, shafts rarely seems to be encountered.

Most fans are driven by electric motors which themselves must be balanced statically and dynamically. Since electrical supplies are commonly of alternating current, vibration must also result from the alternating magnetic forces. These will be at a rate of twice the supply a.c. frequency, since a complete cycle of magnetic attraction occurs for each half cycle of current. Direct current motors may also be subject to some alternating magnetic forces at a frequency dependent on rotational speed and the design of the iron circuits.

Although, when purchased new from the manufacturer, fans are unlikely to give any trouble due to vibration, deterioration in service under arduous conditions may require action to be taken on installation to reduce transmission of subsequent vibration forces. It seems unlikely that such forces will cause structural damage to the buildings in which fans are erected, but are more likely to result first in discomfort or annoyance to the occupants.

4.2. ATTENUATION OF VIBRATION TRANSMISSION

A vibration "isolator" is a form of spring. To appreciate the mechanism of reduction of vibration force transmitted through the spring, a simple system such as that shown in Fig. 4.2 will suffice.

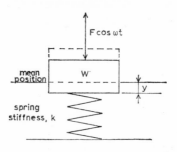

FIG. 4.2. Undamped vibrating system.

The machine, which is subject to a vibration force of $F \cos \omega t$ in a vertical direction, is of weight W and is mounted on an elastic spring whose stiffness is k (that is, the force required to produce unit deflection of the spring, expressed in units of force/length). The system will oscillate about some mean position, and displacement from this position at any instant will be y, considered positive in a downward direction. The additional force exerted by the spring, over and above that due to the weight, will be $k \cdot y$ in an upward direction, or $-k \cdot y$ in a downward direction. The applied force, $F \cos \omega t$, must accelerate the machine mass, and at the same time compress the spring. The equation of motion will be

$$\frac{W}{g} \cdot \frac{\mathrm{d}^2 y}{\mathrm{d} t^2} + k \cdot y = F \cos \omega t \tag{4.4}$$

It is convenient to rewrite this in the form

$$\frac{\mathrm{d}^2 y}{\mathrm{d} t^2} + \frac{kg}{W} \cdot y = \frac{Fg}{W} \cos \omega t \tag{4.5}$$

his differential equation will have a solution in two parts:

(i) the complementary function, which is the solution when the ight-hand side of the equation is zero, that is, when F or ω are zero. his will be of the form

$$y = A \cos \omega_0 t + B \sin \omega_0 t \qquad (4.6)$$

his results in natural vibrations of the system which may be initiated y a displacement of the machine mass, or by a velocity, or both. he value of A in eqn. (4.6) depends on the magnitude of the initial lisplacement, and the value of B on the magnitude of the initial elocity (if any).

(ii) the particular integral due to the form of the applied force, in his case $F \cos \omega t$, which will be

$$y = a \cos \omega t \qquad (4.7)$$

vhere a is known as the amplitude of forced vibration. The full olution of eqn. (4.5), which gives the actual motion of the machine ssembly, will be the sum of eqns. (4.6) and (4.7), thus

$$y = A \cos \omega_0 t + B \sin \omega_0 t + a \cos \omega t \qquad (4.8)$$

Differentiating this equation twice with respect to time and nserting the result in eqn. (4.5) gives

$$-A\omega_0^2 \cos \omega_0 t - B\omega_0^2 \sin \omega_0 t - a\omega^2 \cos \omega t + \frac{kg}{W} A \cos \omega_0 t$$

$$+ \frac{kg}{W} B \sin \omega_0 t + \frac{kg}{W} a \cos \omega t = \frac{Fg}{W} \cos \omega t \qquad (4.9)$$

quating the coefficients of $\cos \omega_0 t$ on each side of the equation,

$$-A\omega_0^2 + \frac{kg}{W} . A = 0$$

hat is,

$$\omega_0 = \sqrt{\frac{kg}{W}} \qquad (4.10)$$

In eqn. (4.10), ω_0 is the angular velocity of natural vibration. The requency of natural vibration f_0 will be $\omega_0/2\pi$, or

$$f_0 = \frac{1}{2\pi} \sqrt{\frac{kg}{W}} \qquad (4.11)$$

Now, $W/k = \delta$, the static deflection of the spring under the weight of the machine, thus

$$f_0 = \frac{1}{2\pi}\sqrt{\frac{g}{\delta}} \qquad (4.12)$$

This shows that the frequency of natural vibration depends only on the static deflection of the spring under the load.

Similarly, the coefficients of $\sin \omega_0 t$ on each side of eqn. (4.9) when equated give

$$-B\omega_0^2 + \frac{kg}{W}.B = 0$$

from which $\omega_0 = \sqrt{(kg/W)}$ as before.

Finally, equating coefficients of $\cos \omega t$ gives

$$-a\omega^2 + \frac{kg}{W}.a = \frac{Fg}{W}$$

$$a(\omega_0^2 - \omega^2) = \frac{F\omega_0^2}{k} \quad \left(\text{since } \omega_0^2 = \frac{kg}{W}\right)$$

from which

$$a = \frac{F}{k}.\frac{1}{1 - (\omega^2/\omega_0^2)} \qquad (4.13)$$

An important fact resulting from eqn. (4.13) is that amplitude a will tend to become infinite as the value of ω approaches the value ω_0. This condition is known as resonance and must be avoided in the design of machine bases.

The result of eqn. (4.8) shows that the effects of natural vibration and of forced vibration occur simultaneously. In practical cases, due to air resistance and energy absorbed by the spring, the effects of natural vibration disappear quite rapidly, leaving only the effect of the forced vibration, thus

$$y = a \cos \omega t$$

where

$$a = \frac{F}{k}.\frac{1}{1 - (\omega^2/\omega_0^2)}$$

It may be seen from eqn. (4.13) that when $\omega > \omega_0$, a is negative, whilst when $\omega < \omega_0$, a is positive. This gives positive and negative values of y for a given value of $F \cos \omega t$. It indicates that, when y is positive, both $F \cos \omega t$ and y are in a downward direction and are said to be in phase, whilst if y is negative, it is in a direction opposite to that of $F \cos \omega t$ and the direction of force and displacement are said to be out of phase by $180°$.

The force transmitted to a structure on which a machine is mounted is the force exerted by the spring on that structure. This will have two components, namely, the force due to the dead weight W, and the force due to the extra compression of the spring due to forced vibration $k.y$. The ratio of the maximum value of the vibration force transmitted $k.a$ to the maximum value of the impressed vibrational force F is known as transmissibility T.

$$T = \frac{ka}{F} = \frac{k}{F} \cdot \frac{F}{k} \cdot \frac{1}{1-(\omega^2/\omega_0^2)}$$

$$= \frac{1}{1-(\omega^2/\omega_0^2)} \tag{4.14}$$

The frequency of forced vibration, $f = \omega/2\pi$, and the frequency of natural vibration, $f_0 = \omega_0/2\pi$, and thus

$$T = \frac{1}{1-(f^2/f_0^2)} \tag{4.15}$$

If a machine base is to reduce the amount of vibration force transmitted, transmissibility must be less than unity, and preferably as little as possible. This can only be achieved if $f > \sqrt{2}.f_0$, when the result from eqn. (4.15) has a negative sign. This sign has no great significance in this context, and for the design of machine bases, it is convenient to rearrange eqn. (4.15):

$$T = \frac{1}{(f^2/f_0^2)-1} \tag{4.16}$$

As a basis for design, it may reasonably be said that f/f_0 should not be less than 2·5, for which $T = 0·19$. Values near to unity will give rise to transmitted forces tending, theoretically, to infinity, with accompanying large amplitudes. Variation of transmissibility T with frequency ratio f/f_0 is shown in the curve $c/c_c = 0$ in Fig. 4.5.

4.3. EFFECT OF VISCOUS DAMPING ON VIBRATION ISOLATION

The solution of the equation of motion, eqn. (4.4), as given by eqn. (4.6) indicates that natural vibration continues for an infinite time when once excited, since it has been assumed that no restraining forces exist. It has been pointed out that such conditions do not exist in practice since there are always restraining forces, however small. Such forces, known as damping forces, may take many forms.

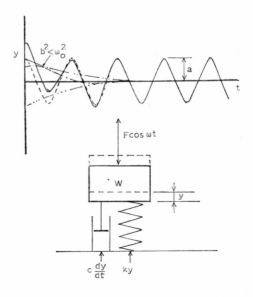

Fig. 4.3. Damped vibrating system.

It is customary to analyse the case where damping forces are due to viscous resistance, and are therefore proportional to velocity since this is probably the simplest mathematical case. Practically, viscous damping may be provided by a piston moving in a cylinder containing viscous fluid, shown diagrammatically in Fig. 4.3. The impressed vibration force is now additionally resisted by the damping force, $c \cdot \mathrm{d}y/\mathrm{d}t$, and eqn. (4.4) may be rewritten

$$\frac{W}{g} \cdot \frac{\mathrm{d}^2 y}{\mathrm{d}t^2} + c \cdot \frac{\mathrm{d}y}{\mathrm{d}t} + ky = F \cos \omega t \qquad (4.17)$$

or

$$\frac{d^2y}{dt^2} + 2b \cdot \frac{dy}{dt} + \omega_0^2 y = \frac{F\omega_0^2}{k} \cdot \cos \omega t \qquad (4.18)$$

where $2b = cg/W$ and $\omega_0^2 = kg/W$.

As with eqn. (4.4) (or eqn. (4.5)) the value of y satisfying eqns. (4.17) and (4.18) may be expressed in two parts;

(i) the complementary function when $F \cos \omega t = 0$ will be of the form

$$y = C \exp(rt) \qquad (4.19)$$

and thus, $dy/dt = Cr \exp(rt)$ and $d^2y/dt^2 = Cr^2 \exp(rt)$. On substituting these values into eqn. (4.18) with $F \cos \omega t = 0$ gives $(r^2 + 2br + \omega_0^2)C \exp(rt) = 0$, from which

$$r = -b \pm \sqrt{(b^2 - \omega_0^2)} \qquad (4.20)$$

If $b^2 > \omega_0^2$, the value of r is real and negative and the solution, $y = C \exp(-rt)$ represents a displacement initially of magnitude C slowly reducing to zero.

If $b^2 < \omega_0^2$, the value of r has an imaginary part, thus

$$r = -b \pm j\omega_1, \text{ where } \omega_1^2 = \omega_0^2 - b^2$$

and

$$y = \exp(-bt) \cdot (C_1 \cos \omega_1 t + C_2 \sin \omega_1 t) \qquad (4.21)$$

In this case vibration does occur, but the amplitude will decrease logarithmically to zero.

In either case y has only a transient value, reducing to zero quite rapidly in practice. The changeover from non-oscillatory to oscillatory motion occurs when $\omega_0 = b$, that is, when $\sqrt{(kg/W)} = cg/2W$. The critical value of c for this condition may be written c_c, and has the value

$$c_c = 2\sqrt{\frac{kW}{g}} \qquad (4.22)$$

(ii) The particular integral for the impressed force $F \cos \omega t$ will have the form

$$y = a \cos(\omega t + \phi) \qquad (4.23)$$

where a is the amplitude of forced vibration and ϕ is a phase angle indicating that the displacement, although vibrating with the same pulsatance (angular velocity) as the impressed force, is out of step with it.

Substituting $dy/dt = -a\omega \sin(\omega t + \phi)$ and
$$d^2y/dt^2 = -a\omega^2 \cos(\omega t + \phi)$$

in eqn. (4.18) gives

$$-a\omega^2 \cos(\omega t + \phi) - 2ba\omega \sin(\omega t + \phi) + \omega_0^2 a \cos(\omega t + \phi)$$
$$= \frac{F\omega_0^2}{k}\cos\omega t \qquad (4.24)$$

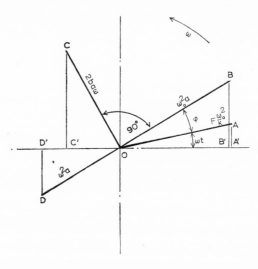

FIG. 4.4. Equivalent rotating vectors for a damped vibrating system.

Equation (4.21) may be represented by a rotating vector system, having an angular velocity of ω, as shown in Fig. 4.4. In this diagram vector **OA** has a length of $F\omega_0^2/k$, **OB** of $\omega_0^2 a$, **OC** of $2ba\omega$ and **OD** of $\omega^2 a$. The intercepts OA', OB', OC' and OD' represent the terms of eqn. (4.24). Resolving in directions OB and OC,

$$\frac{F\omega_0^2}{k} . \cos\phi = \omega_0^2 a - \omega^2 a$$

$$\frac{F\omega_0^2}{k} . \sin\phi = -2ba\omega$$

and, eliminating ϕ,

$$\left(\frac{F\omega_0^2}{k}\right)^2 (\cos^2 \phi + \sin^2 \phi) = a^2(\omega_0^2 - \omega^2)^2 + (2ba\omega)^2$$

from which

$$a = \frac{F\omega_0^2}{k\sqrt{[(\omega_0^2 - \omega^2)^2 + (2b\omega)^2]}}$$

$$= \frac{F}{k\sqrt{[(1 - \omega^2/\omega_0^2)^2 + (2b/\omega_0 \cdot \omega/\omega_0)^2]}}$$

$$= \frac{F}{k\sqrt{[(1 - \omega^2/\omega_0^2)^2 + (2c/c_c \cdot \omega/\omega_0)^2]}} \quad (4.25)$$

since

$$\frac{2b}{\omega_0} = \frac{cg}{W}\sqrt{\frac{W}{kg}} = c\sqrt{\frac{g}{Wk}} = \frac{2c}{c_c}.$$

It is interesting to note that amplitude a no longer has an infinite value when $\omega = \omega_0$.

To find transmissibility it is necessary to find the force exerted on the structure, on which the isolating base is mounted, due to vibration. This consists of two parts: that due to the spring force, $ky = ka \cos (\omega t + \phi)$, and that due to the reaction of the damping force, $c \cdot dy/dt = -ca\omega \sin (\omega t + \phi)$. These two forces are vectorially at right angles and may be combined to give a maximum value of

$$\sqrt{(ka)^2 + (ca\omega)^2} = ka\sqrt{1 + (c\omega/k)^2}$$

$$= ka\sqrt{1 + \left(\frac{2c}{c_c} \cdot \frac{\omega}{\omega_0}\right)^2} \quad (4.26)$$

since

$$\frac{c\omega}{k} = \frac{c\omega}{\sqrt{k}\sqrt{k}} = \frac{c\omega}{\omega_0}\sqrt{\frac{g}{Wk}} = \frac{2c}{c_c} \cdot \frac{\omega}{\omega_0}$$

Thus transmissibility becomes

$$T = \frac{ka}{F} \sqrt{1 + \left(\frac{2c}{c_c} \cdot \frac{\omega}{\omega_0}\right)^2}$$

$$T = \frac{\sqrt{1 + \left(\frac{2c}{c_c} \cdot \frac{\omega}{\omega_0}\right)^2}}{\sqrt{(1 - \omega^2/\omega_0^2)^2 + \left(\frac{2c}{c_c} \cdot \frac{\omega}{\omega_0}\right)^2}}$$

$$= \frac{\sqrt{1 + \left(\frac{2c}{c_c} \cdot \frac{f}{f_0}\right)^2}}{\sqrt{(1 - f^2/f_0^2)^2 + \left(\frac{2c}{c_c} \cdot \frac{f}{f_0}\right)^2}} \tag{4.27}$$

Not only does amplitude a of forced vibration have a finite value at resonance, but also does transmissibility. However, in the working range of vibration isolators, the effect of damping is to reduce the effectiveness of the isolator for values of $f/f_0 > \sqrt{2}$, since T has a greater value than that given by eqn. (4.15) (to which eqn. (4.27) reduces on putting $c = 0$). There are circumstances, such as with fans having large heavy impellers which take a long time to run up to speed on starting, where a moderate value of c/c_c (0·1–0·2) will reduce the transient vibration amplitude as f increases through the value of f_0.

It is often convenient to express the effectiveness of a vibration isolator in terms of attenuation in decibels (dB), thus

$$\text{attenuation} = 20 \log_{10} \frac{1}{\text{transmissibility}} \text{ dB} \tag{4.28}$$

Alternatively, the degree of isolation, expressed as a percentage, may be quoted:

$$\% \text{ isolation} = (1 - \text{transmissibility}) \times 100\% \tag{4.29}$$

4.4. OTHER MODES OF VIBRATION

The theory so far outlined assumes a point load with a periodic force acting along the axis of a spring. The system comprising a machine mounted on vibration isolators is rarely such a simple system. A number of units may be placed at suitable points beneath the base of the machine, whose centre of gravity may not be disposed symmetrically. Not only is a vertical translatory motion possible, ut also translatory motions in two horizontal directions at right ingles, and, in addition, torsional motions in one horizontal and two vertical planes. There are thus six forms, known as modes, of vibration. Frequencies of natural vibration may become more difficult to calculate. Under certain conditions there may be separate

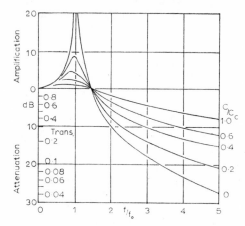

FIG. 4.5. Attenuation by vibration isolators.

natural frequencies for each mode, whilst for others there may be a single combined, or " coupled ", complex natural vibration. A full appreciation of the three-dimensional problem involves the simultaneous solution of six equations of motion. The process may be

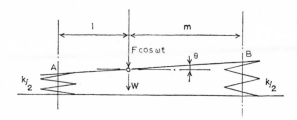

FIG. 4.6. Undamped vibrations in two modes.

illustrated by considering a comparatively simple case of two modes in a single plane, one being in translation and the other in torsion (Fig. 4.6). An impressed force of $F \cos \omega t$ is applied to the centre of gravity of the load (or machine) W, unsymmetrically disposed with respect to the spring mountings, which are each taken to have half the total stiffness k. Consequently, as the load vibrates, it will have a vertical translatory motion y causing a rotational motion about the centre of gravity due to a net torque of $k(m-l)y$. As the load rotates about the centre of gravity by an angle θ, it will cause a vertical movement of the centre of gravity of $(m-l)\theta$. Damping forces are taken as being very small. The equation of motion corresponding to eqn. (4.4) becomes

$$\frac{W}{g}\cdot\frac{d^2y}{dt^2}+ky-\frac{k}{2}(m-l)\theta = F\cos\omega t \qquad (4.30)$$

It is seen that, due to the proportion of m to l, the net vertical movement due to torsion is upwards, resulting in a vertical force in the opposite direction to that due to the translatory displacement. Similarly, in torsion,

$$\frac{W}{g}\cdot r^2\cdot\frac{d^2\theta}{dt^2}+\frac{k}{2}(y+l\theta)l-\frac{k}{2}(y-m\theta)m = 0 \qquad (4.31)$$

where r is the radius of gyration of the load about its centre of gravity. Neglecting transient natural vibrations, particular solutions of eqns. (4.30) and (4.31) due to the impressed force, $F\cos\omega t$, will be

$$y = a\cos\omega t \qquad (4.32)$$

$$\theta = \theta_0\cos\omega t \qquad (4.33)$$

Substituting these values in eqns. (4.30) and (4.31) gives

$$a\left(k-\frac{W}{g}\omega^2\right)-\frac{k}{2}(m-l)\theta_0 = F$$

$$\theta_0\left[\frac{k}{2}(m^2+l^2)-\frac{W}{g}r^2\omega^2\right]-\frac{k}{2}(m-l)a = 0$$

and eliminating θ_0 gives

$$ka\left(1-\frac{W}{kg}\omega^2\right)\left[\frac{k}{2}(m^2+l^2)-\frac{W}{g}\cdot r^2\omega^2\right]-\frac{k^2a}{4}(m-l)^2$$

$$= F\left[\frac{k}{2}(m^2+l^2)-\frac{W}{g}\cdot r^2\omega^2\right]$$

$$k^2a\left[\left(1-\frac{\omega^2}{\omega_0^2}\right)\left(\frac{m^2+l^2}{2}-\frac{\omega^2}{\omega_0^2}\cdot r^2\right)-\frac{(m-l)^2}{4}\right]=Fk\left[\frac{m^2+l^2}{2}-\frac{\omega^2}{\omega_0^2}\cdot r^2\right]$$

$$kar^2\left[\left(1-\frac{\omega^2}{\omega_0^2}\right)\left(\frac{m^2+l^2}{2r^2}-\frac{\omega^2}{\omega_0^2}\right)-\frac{(m-l)^2}{4r^2}\right]=Fr^2\left[\frac{m^2+l^2}{2r^2}-\frac{\omega^2}{\omega_0^2}\right]$$

$$ka\left[\frac{\omega^4}{\omega_0^4}-\left(\frac{m^2+l^2}{2r^2}+1\right)\frac{\omega^2}{\omega_0^2}+\frac{(m+l)^2}{4r^2}\right]=F\left[\frac{m^2+l^2}{2r^2}-\frac{\omega^2}{\omega_0^2}\right]$$

$$a=\frac{F}{k}\cdot\frac{[(m^2+l^2)/2r^2]-(\omega^2/\omega_0^2)}{(\omega^4/\omega_0^4)-\{[(m^2+l^2)/2r^2]+1\}(\omega^2/\omega_0^2)+[(m+l)^2/4r^2]}$$

$$(4.34)$$

Similarly, eliminating a instead of θ_0 gives

$$\theta_0=\frac{F}{k}\cdot\frac{(m^2-l^2)/2r^2}{(\omega^4/\omega_0^4)-\{[(m^2+l^2)/2r^2]+1\}(\omega^2/\omega_0^2)+[(m+l)^2/4r^2]}$$

$$(4.35)$$

When the denominator of either of eqns. (4.34) or (4.35) becomes zero, the amplitudes a and θ_0 become infinite and there is a condition of resonance. The frequency at which this occurs is the frequency of natural vibration and is not generally $\omega_0/2\pi$ as would be expected for translational vibration alone. If $(m+l)^2 = 4r^2$ and $m^2+l^2=2r^2$, that is, if $ml = r^2$,

$$a=\frac{F}{k}\frac{(1-\omega^2/\omega_0^2)}{(1-\omega^2/\omega_0^2)^2}$$

$$=\frac{F}{k}\cdot\frac{1}{(1-\omega^2/\omega_0^2)}$$

which is identical to eqn. (4.13). Under these circumstances there is no coupling.

4.5. PRACTICAL MOUNTING DESIGN

Since the inherent damping in most materials used as vibration isolating materials is of a low order ($c/c_c < 0.2$ in general) it is probably sufficient to use the simple expression for transmissibility of eqn. (4.16),

$$T = \frac{1}{f^2/f_0^2 - 1}$$

If $f/f_0 > 4$, this may be simplified still further to

$$T = \frac{f_0^2}{f^2} \qquad (4.36)$$

to a sufficient degree of accuracy, since there is often difficulty in predicting the natural frequency f_0.

Materials such as cork, rubber and felt behave as if they have greater stiffness under alternating forces than under steady forces. Rubber and cork have dynamic stiffness of the order of twice the static stiffness, depending on loading, shape and frequency, whilst for felt this ratio may be rather higher. It is thus necessary to know the true natural frequency (or resonant frequency) of the material or isolator under the conditions of use. If this is not directly available the " dynamic " modulus of elasticity for the relevant conditions may be known, the dynamic modulus E_D being defined by

$$E_D = \frac{\text{alternating stress}}{\text{strain}} \qquad (4.37)$$

where the alternating stress has an average value equal to the load divided by the area of material. An equivalent " static " deflection δ_D for use in eqn. (4.12) to find frequency of natural vibration f_0 is then found:

$$\delta_D = \frac{\text{load} \times \text{thickness}}{\text{area} \times E_D} \qquad (4.38)$$

For convenient use, eqn. (4.12) becomes

$$f_0 = 3.13/\sqrt{\delta} \text{ c/s} \qquad (4.39)$$

where δ is in inches, or

$$f_0 = 15.8/\sqrt{\delta} \text{ c/s} \qquad (4.40)$$

where δ is in millimetres.

To avoid coupling of natural frequencies, it is desirable to have the upper side of the vibration isolators in contact with the machine in the plane of the centre of gravity of the machine assembly, and to ensure that, when the load is lowered on to the isolators, the deflection is purely vertical and without any torsional component. Any connections, such as electrical connections to a motor, air duct connections and pipes, should have flexible sections of lower stiffness than the isolators in any direction.

Clearance for vibration amplitudes during starting periods is desirable. For rotational out-of-balance, the vibration force F is proportional to (speed of rotation)^2 as shown in eqn. (4.1). From eqn. (4.13) amplitude of vibration a becomes

$$a = \frac{M\omega^2}{gk} \bigg/ (1 - \omega^2/\omega_0^2)$$

$$= \frac{M\omega_0^2}{gk} \cdot \frac{\omega^2/\omega_0^2}{1 - \omega^2/\omega_0^2} \qquad (4.41)$$

When $\omega = \omega_0$, that is at resonance, the amplitude tends to have an infinite value. This is likely to occur when a machine runs up to speed (or slows down) and it is sometimes advisable to fix rubber blocks, known as " snubbers ", to restrict movement. When $\omega \gg \omega_0$ the amplitude tends to a constant value of $M\omega^2/gk$. At the same time, the value of the transmitted force tends to a constant value, the impressed force increasing at the same rate as the transmissibility is decreasing.

If fans are mounted on a heavy rigid floor, it is probably unnecessary to use vibration isolators since the floor will be so massive that vibration accelerations will be negligible and will not be sensed.

4.6. SOUND WAVES IN AIR

Sound may be defined as a disturbance in an elastic medium capable of exciting the sense of hearing. Sound in air is a minute disturbance of the ambient atmospheric pressure which travels from the source of the sound through the air to the ear. The air itself does not flow, but " particles " of air oscillate as the disturbance passes in a similar way to that of a float which " bobs " on the surface of a pond at the passage of a ripple.

If the disturbance could be observed as it was about to leave the source its profile could perhaps be seen to be capable of representation by some mathematical expression, $\phi = f(X)$, where ϕ is some defined property of the disturbance such as pressure, for example, and X is some distance from a datum on the disturbance. If this disturbance travels through the air unchanged with a velocity c, after time t its local datum will have travelled a distance ct. If any point within the disturbance is a distance x from the source it is at a

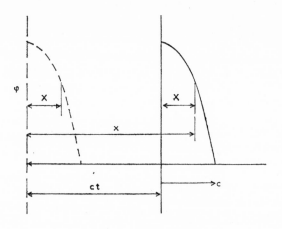

Fig. 4.7. Progressive wave.

distance $X = x - ct$ from the local datum (Fig. 4.7); thus, with the source as datum,

$$\phi = f(X) = f(x - ct) \tag{4.42}$$

The simplest form of disturbance is a harmonic wave, for example

$$\phi = \phi_0 \cos(x - ct) \tag{4.43}$$

from which

$$\frac{d^2\phi}{dx^2} = -\phi_0 \cos(x - ct)$$

and

$$\frac{d^2\phi}{dt^2} = -c^2\phi_0 \cos(x - ct)$$

giving

$$\frac{d^2\phi}{dt^2} = c^2 \cdot \frac{d^2\phi}{dx^2} \tag{4.44}$$

Equation (4.44) is known as the equation of wave motion. It will be seen that the function $\phi = \phi_0 \cos(ct - x)$ also satisfies this equation.

A wave passing unchanged in this manner is known as a plane wave. Most sound waves are three-dimensional and for these eqn. (4.44) must be rewritten

$$\frac{d^2\phi}{dt^2} = c^2 \nabla^2 \phi \tag{4.45}$$

where $\nabla^2 \phi$ is the second derivative of ϕ with respect to three co-ordinate dimensions. Nevertheless, many of the basic acoustic relationships may be derived from consideration of plane waves.

It remains to be established that a disturbance in air can give rise to the progressive wave whose form is described above. Consider an elemental prism of air of cross-sectional area S, and length dx, to be subjected to a disturbance in the direction of x only which results in one boundary being displaced by ξ and the other by $\xi + (d\xi/dx).dx$. The length of the element will now be $dx + (d\xi/dx).dx = dx[1 + (d\xi/dx)]$. The mass of the element will remain unchanged, so

$$\rho_0 S.dx = \rho.S.dx\left(1 + \frac{d\xi}{dx}\right)$$

where ρ_0 is the undisturbed density of the air and ρ is the density after disturbance.

$$\rho = \rho_0\left(1 + \frac{d\xi}{dx}\right)^{-1} = \rho_0\left(1 - \frac{d\xi}{dx}\right) \tag{4.46}$$

since $d\xi/dx$ in practice is very small, being of the order of 10^{-4} for very intense sounds which are painful to the ear. Differentiating eqn. (4.46),

$$\frac{d\rho}{dx} = -\rho_0 \frac{d^2\xi}{dx^2} \tag{4.47}$$

The net force in the direction of $x = pS - (p + dp)S = -dp.S$. Thus the equation of motion will be

$$-dp.S = \rho_0 S.dx.\frac{d^2\xi}{dt^2}$$

or

$$\frac{dp}{dx} = -\rho_0.\frac{d^2\xi}{dt^2} \tag{4.48}$$

Now, dp/dx may be written $dp/d\rho . d\rho/dx$, and substituting this and eqn. (4.47) in eqn. (4.48) gives

$$-\frac{dp}{d\rho} . \rho_0 . \frac{d^2\xi}{dx^2} = -\rho_0 . \frac{d^2\xi}{dt^2}$$

or

$$\frac{d^2\xi}{dt^2} = \frac{dp}{d\rho} . \frac{d^2\xi}{dx^2} = c^2 . \frac{d^2\xi}{dx^2} \qquad (4.49)$$

where

$$c^2 = dp/d\rho \qquad (4.50)$$

and will be seen to have the dimensions of (velocity)^2. Equation (4.49) is identical in form to eqn. (4.44) and thus it is shown that a disturbance in air (or other similar fluid) can give rise to wave motion.

The period of an individual disturbance constituting sound may be from 1/20 to 1/20,000 second in duration, with the result that it is unlikely that any heat transfer to the surrounding fluid will take place in such a short time. Thus the process of compression and rarefaction may reasonably be regarded as adiabatic, and $pV^\gamma = $ constant, V being a volume element. Now the mass of the element, ρV, remains constant and consequently $p\rho^{-\gamma} = $ constant. Differentiating,

$$dp . \rho^{-\gamma} - p . \gamma . \rho^{-\gamma-1} . d\rho = 0$$

from which

$$\frac{dp}{d\rho} = \frac{\gamma p}{\rho}$$

Equation (4.50) shows that

$$c^2 = \frac{dp}{d\rho} = \frac{\gamma p}{\rho}$$

Thus the velocity of sound in air will be

$$c = \sqrt{\frac{\gamma p}{\rho}} \qquad (4.51)$$

where p is the ambient atmospheric pressure and ρ the air density. Since, by the gas laws (eqn. (1.1)) $\rho = p/RT$

$$c = \sqrt{(\gamma RT)} \qquad (4.52)$$

For air, $\gamma = 1\cdot41$ and $R = 53\cdot3$ ft lbf/lb °R, so

$$c = \sqrt{(1\cdot41 \times 53\cdot3 \times 32\cdot2 \times T)}$$

$$= 49\cdot2\sqrt{T} \text{ ft/s.} \tag{4.53}$$

If the air temperature is 68°F (20°C)

$$c = 49\cdot2\sqrt{(460+68)} = 1131 \text{ ft/s.}$$

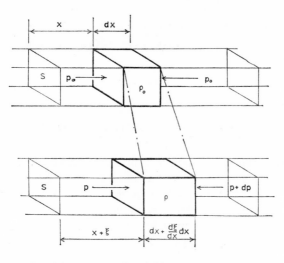

Fig. 4.8. Element of air subject to a disturbance.

In metric units $R = 29\cdot26$ kg m/kg °K, so

$$c = \sqrt{(1\cdot41 \times 29\cdot26 \times 9\cdot81 \times T)}$$

$$= 20\cdot1\sqrt{T} \text{ m/s.} \tag{4.53a}$$

If the air temperature is 20°C,

$$c = 20\cdot1\sqrt{(273+20)} = 344 \text{ m/s.}$$

These values agree quite well with experimentally determined values.

The form of particle displacement ξ which results from eqn. (4.49) is generally written

$$\xi = \xi_a \cos(\omega t - kx) + \xi_b \cos(\omega t + kx) \tag{4.54}$$

$$= \xi_a \cos k(ct - x) + \xi_b \cos k(ct + x) \tag{4.55}$$

where $k = \omega/c$ and is known as the wave number. The first part of eqn. (4.55) is seen to be of the same form as that of eqn. (4.43) and represents a plane wave leaving a sound source. The second part of eqn. (4.55) is similar to the first but with the sign of x reversed. It thus represents a plane wave returning to the sound source (as, for example, a reflected wave). The magnitude of ξ is repeated at intervals of distance $x = \lambda$, such that $k\lambda = 2\pi$, at any instant of time. Wavelength $\lambda = 2\pi/k = 2\pi c/\omega$. Similarly, at any point, the magnitude of ξ repeats itself at intervals of time T such that $\omega T = 2\pi$, and therefore, $T = 2\pi/\omega$. Since T is the time taken for a single complete disturbance, the number of complete disturbances in unit time, the frequency f is $\omega/2\pi$. Thus

$$f . \lambda = c \tag{4.56}$$

4.7. SOUND PRESSURE, PARTICLE VELOCITY AND SOUND INTENSITY

The instantaneous change in air pressure p due to the passage of a sound wave may be expressed as the difference between the atmospheric pressure at any instant, p, and the mean value p_0. The loudest bearable sound would be represented by a sound pressure of the order of one five-thousandth of the atmospheric pressure and consequently p may be regarded as being identical with $\mathrm{d}p$. Thus

$$p = \mathrm{d}p = \frac{\mathrm{d}p}{\mathrm{d}\rho} . \mathrm{d}\rho = c^2 \mathrm{d}\rho$$

Similarly,

$$\mathrm{d}\rho = \rho - \rho_0 = -\rho_0 . \frac{\mathrm{d}\xi}{\mathrm{d}x}$$

(from eqn. (4.46)).
Thus sound pressure

$$p = -c^2 \rho_0 . \frac{\mathrm{d}\xi}{\mathrm{d}x} \tag{4.57}$$

For a simple plane wave leaving a source and not being reflected back, eqn. (4.54) may be substituted in eqn. (4.57), and

$$p = -c^2 \rho_0 k \xi_0 \sin(\omega t - kx) \tag{4.58}$$

Particle velocity

$$u = \frac{d\xi}{dt} \tag{4.59}$$

and for a simple outgoing plane wave

$$u = -\xi_0 \omega \sin(\omega t - kx) \tag{4.60}$$

For this simple form of plane wave propagation it can be seen that

$$\frac{p}{u} = \frac{c^2 \rho_0 k}{\omega} = \rho_0 c \tag{4.61}$$

Referring to eqn. (2.1) it can be seen that air power is the product of pressure difference and air volume flow. The product of pressure difference and air velocity will be the air power per unit area of cross-section at right angles to the direction of flow. The sound power per unit area will be the product of sound pressure and particle velocity, that is $p \times u$. For simple outgoing plane waves,

$$p \times u = \frac{p^2}{\rho_0 c} = \frac{p^2 \sin^2(\omega t - kx)}{\rho_0 c}$$

The time average value of this expression is known as the sound intensity I and since the average value of $\sin^2 \theta$ is $\frac{1}{2}$,

$$I = \frac{p^2}{2\rho_0 c} = \frac{p_{\text{r.m.s.}}^2}{\rho_0 c} \tag{4.62}$$

where $p_{\text{r.m.s.}}$ is known as the root mean square value of p.

Sound pressure is expressed in units of newton/metre^2 (N/m^2) (dyne/cm$^2 \times 10$), one newton being the force which will accelerate a mass of one kilogramme by one metre per second per second. This gives a unit of energy of one newton-metre, or one joule, and a unit of power of one joule per second, which is watt. For air at a temperature of 20°C and a density of 1·2 kg/m^3, a r.m.s. sound pressure of 2×10^{-5} N/m^2 (0·0002 dyne/cm^2) gives rise to an intensity in a plane outgoing wave of

$$
\begin{aligned}
I &= \frac{p_{\text{r.m.s.}}^2}{\rho_0 c} = \frac{p_{\text{r.m.s.}}^2}{\rho_0 . 20 \cdot 1 \sqrt{T}} \\
&= \frac{(2 \times 10^{-5})^2}{1 \cdot 2 \times 20 \cdot 1 \times \sqrt{293}} = 0 \cdot 97 \times 10^{-12} \text{ watt/m}^2
\end{aligned} \tag{4.63}
$$

A plane wave of cross-sectional area S will traverse a volume of air of $c.S$ units in unit time. Thus sound energy per unit volume, known as sound energy density,

$$E = \frac{I.S}{c.S} = \frac{I}{c} = \frac{p_{r.m.s.}^2}{\rho_0 c^2} \qquad (4.64)$$

4.8. LEVELS

Acoustical quantities, power, pressure and intensity, are usually minute when expressed in normal m.k.s. units. Moreover, the range of values for which the ear is sensitive is very large. For example, the sound pressure which causes acute discomfort (threshold of painful hearing) is about one million times as great as that which can just be heard under the most favourable circumstances (threshold of audibility). It has been found convenient to express the physical quantities in terms of a logarithmic scale of power ratios, and to refer to the difference in " level " between any two positions on the scale. For example, the difference in level between power W_2 and power W_1 is said to be $10 \log_{10}(W_2/W_1)$ decibels (dB). A scale of power level results from allotting an agreed value of 10^{-12} watt to W_1. Thus sound power level L_w is given by

$$L_w = 10 \log_{10} \frac{W \text{ watt}}{10^{-12}} \text{ dB} \qquad (4.65)$$

Since sound intensity is sound power per unit area, sound intensity level L_i may be expressed as

$$L_i = 10 \log_{10} \frac{I \text{ watt/m}^2}{10^{-12}} \text{ dB} \qquad (4.66)$$

In eqn. (4.65), a reference sound power of 10^{-12} watt has been selected and this naturally leads to a reference sound intensity of 10^{-12} watt/m^2 (10^{-16} watt/cm^2). There is reasonable international agreement on these values.

In view of the relationship for plane progressive waves given by eqn. (4.62), namely that $I \propto p^2$, sound pressure level, L_p, may be expressed as:

$$L_p = 10 \log_{10} \left(\frac{p_{r.m.s.}^2 \text{ N/m}^2}{0 \cdot 00002} \right)^2 \text{ dB} \qquad (4.67)$$

t can be seen from the relationship expressed in eqn. (4.63) that
under normal air conditions a plane wave intensity of 10^{-12} watt/m^2
corresponds very nearly to a r.m.s. sound pressure of 0·00002 N/m^2,
and under such circumstances, sound intensity level and sound pressure level are numerically the same. These conditions, for which
here is no reflection back of the outgoing sound wave, are known as
free field conditions. They are encountered very nearly in practice
in large unbounded spaces, or rooms having fully absorptive surfaces.

4.9. ADDITION OF SOUND WAVES

Considering a point in a free field subject to sound pressure from
two different sources, the resulting sound pressure at any instant will
be the sum of the two pressures, for example;

$$p_{a+b} = p_a \cos (k_a x_a - \omega_a t) + p_b \cos (k_b x_b - \omega_b t + \phi)$$

where ϕ is an angular difference in phase. It follows that

$$p_{a+b}^2 = p_a^2 \cos^2 (k_a x_a - \omega_a t) + p_b^2 \cos^2 (k_b x_b - \omega_b t + \phi)$$
$$+ 2 p_a p_b \cos (k_a x_a - \omega_a t)[\cos (k_b x_b - \omega_b t) \cos \phi$$
$$- \sin (k_b x_b - \omega_b t) \sin \phi]$$

This will have a mean square value of

$$p_{A+B}^2 = \tfrac{1}{2} p_a^2 + \tfrac{1}{2} p_b^2; \quad \cos (k_a x_a - \omega_a t) \neq \cos (k_b x_b - \omega_b t) \qquad (4.68)$$

$$p_{A+B}^2 = \tfrac{1}{2} p_a^2 + \tfrac{1}{2} p_b^2 + p_a p_b \cos (2kx + \phi),$$
$$\cos (k_a x_a - \omega_a t) = \cos (k_b x_b - \omega_b t) \qquad (4.69)$$

The case stated by eqn. (4.69) is a case unlikely to be encountered in
noise control, where sound disturbances are usually random, thus,
generally,

$$p_{A+B}^2 = p_A^2 + p_B^2 \qquad (4.70)$$

that is, the total mean square pressure at a point is the sum of the
individual mean square pressures. In eqn. (4.70), p_{A+B}, p_A and p_B
are r.m.s. pressures.

If the sound pressure levels of these individual sound pressures are
A and B respectively, and the level due to both together is C, then

$$A = 10 \log_{10} \frac{p_A^2}{p_0^2} \quad \text{and} \quad B = 10 \log_{10} \frac{p_B^2}{p_0^2}$$

H

where p_0 is the reference r.m.s. pressure. Thus

$$p_A^2 = p_0^2 \log_{10}^{-1} \frac{A}{10} \quad \text{and} \quad p_B^2 = p_0^2 \log_{10}^{-1} \frac{B}{10}$$

$$p_{A+B}^2 = p_A^2 + p_B^2 = p_0^2 \left(\log_{10}^{-1} \frac{A}{10} + \log_{10}^{-1} \frac{B}{10} \right)$$

and

$$C = 10 \log_{10} \frac{p_{A+B}^2}{p_0^2} = 10 \log_{10} \left(\log_{10}^{-1} \frac{A}{10} + \log_{10}^{-1} \frac{B}{10} \right) \text{ dB}$$

(4.71)

Similarly, subtractions of sound pressure may be performed since

$$p_A^2 = p_{A+B}^2 - p_B^2$$

and

$$A = 10 \log_{10} \left(\log_{10}^{-1} \frac{C}{10} - \log_{10}^{-1} \frac{B}{10} \right) \text{ dB}$$

(4.72)

Equation (4.72) is particularly useful when sound pressure level due to an equipment is measured in a space whose background sound pressure level is very little less than the measured level. For example, if the measured level due to both background and equipment is 73 dB and due to the background alone is 70 dB, the level due to the equipment which would have been measured in the absence of any background noise would be

$$10 \log_{10} (\log_{10}^{-1} 7 \cdot 3 - \log_{10}^{-1} 7 \cdot 0) \text{ dB}$$

$$= 10 \log_{10}(2 \times 10^7 - 1 \times 10^7) = 10 \log_{10} 10^7 = 70 \text{ dB}$$

Sound intensity levels and sound power levels may be dealt with similarly.

4.10. DIFFUSE SOUND FIELDS IN LARGE ROOMS

If a source of sound is operated continuously in a large enclosure the sound energy in the enclosure will increase until the rate at which the sound energy is absorbed by the medium and room surfaces is equal to the rate at which the sound energy is generated by the source. If the source suddenly stops emitting sound energy, there will be a period of time (theoretically infinite) before the sound energy in the room is entirely absorbed. In normal sized rooms, the amount of sound energy absorbed by the air will generally be very small compared with that absorbed by the room surfaces. Defining the absorption coefficient α as being the ratio of sound energy absorbed by unit area of a surface to the sound energy incident on that unit surface area, this may be taken as having an average value $\bar{\alpha}$ for a number of different surfaces of areas $S_1, S_2, \ldots, S_n$, of

$$\bar{\alpha} = \frac{\Sigma S\alpha}{\Sigma S} \tag{4.73}$$

Suppose the steady energy density in a large room of volume V is homogeneous throughout the space and has a value of E_0, and that

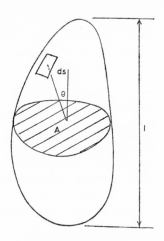

FIG. 4.9. Bounded space.

the source is suddenly cut off. After the first complete reflection from the room surfaces, the total sound energy left in the room will be $E_1V = E_0V - E_0V.\bar{\alpha} = E_0V(1-\bar{\alpha})$. After n complete reflections, the energy remaining will be

$$E_nV = E_0V(1-\bar{\alpha})^n \tag{4.74}$$

Now if t is the time taken for n complete reflections and L is the mean length of path between reflections,

$$n = \frac{ct}{L} \tag{4.75}$$

To find the mean length of path, L, a three-dimensional space as shown in Fig. 4.9 is considered. In any one direction, the volume of the space, $V = \int l.dA = \bar{l}.A$, where $\bar{l}$ is the average length of l. This relationship will be true in all directions in the space, and the mean free path will have a length L which is the mean value of l in all directions, that is

$$L = \frac{\sum\limits_{n=1}^{\infty} l_n A_n}{\sum\limits_{n=1}^{\infty} A_n}$$

and, since $l_n A_u = V$ for all values of n

$$L = \frac{V}{\bar{A}} \tag{4.76}$$

Now area A is half the projection of the surface area S since S covers A twice (once from each side) and thus

$$A = \tfrac{1}{2} \int ds \, . \cos \theta$$

Averaging this over all directions gives

$$\bar{A} = \tfrac{1}{2} \int \overline{\cos \theta} \, ds \tag{4.77}$$

where $\cos \theta$ is the angle between the normal to A and the normal to ds, and $\overline{\cos \theta}$ is the average value which may be found by reference to Fig. 4.10. The area of a zone of a hemispherical surface is $r \, . d\theta \times 2\pi r . \sin \theta$, and

$$\overline{\cos \theta} = \frac{\Sigma \cos \theta \times \text{area of zone}}{\text{total area of surface}}$$

$$= \int\limits_0^{\pi/2} \frac{2\pi r^2 \sin \theta . \cos \theta . d\theta}{2\pi r^2}$$

$$= \tfrac{1}{2} \int\limits_0^{\pi/2} \sin 2\theta . d\theta = -\tfrac{1}{4} [\cos 2\theta]_0^{\pi/2} = \tfrac{1}{2}$$

Inserting this result in eqn. (4.77) gives $\bar{A} = S/4$, and from eqn. (4.76)

$$L = \frac{4V}{S} \tag{4.78}$$

and substituting this result in eqns. (4.75) and (4.74):

$$E_n = E_0 (1 - \bar{\alpha})^{cSt/4V}$$

Now reverberation time T of an enclosure is defined as the time taken for the sound energy to decay by 60 dB, thus

$$-60 = 10 \log_{10} \frac{E_n}{E_0} = 10 \log_{10}(1-\bar{\alpha})^{cST/4V}$$

or

$$\frac{24V}{cST} = -\log_{10}(1-\bar{\alpha})$$

$$T = \frac{-24V}{cS \cdot \log_{10}(1-\bar{\alpha})} \tag{4.79}$$

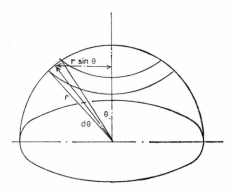

Fig. 4.10. Element of hemispherical space.

Equation (4.79) may also be written

$$T = \frac{-24 \cdot 2 \cdot 303V}{cS \cdot \log_e(1-\bar{\alpha})} = \frac{-55 \cdot 3V}{cS \cdot \log_e(1-\bar{\alpha})}$$

If $\bar{\alpha}$ is very small, $\log_e(1-\bar{\alpha})$ is very nearly the same as $-\bar{\alpha}$, thus

$$T = \frac{55 \cdot 3V}{cS\bar{\alpha}} \tag{4.80}$$

Equation (4.79) or (4.80) may be used to compute the value of $\bar{\alpha}$ from measurements of reverberation time.

If a source emits sound into a space at a steady rate, the resulting sound field may be regarded as being in two parts. One is the direct field due to sound radiating directly from the source; the other is the field associated with reflection from the room surfaces and is known as the reverberant field.

Considering first the direct field, the intensity at any distance from a point source of sound power W, situated in free space, will b $I = power/area = W/4\pi d^2$. If the source is in the plane of a infinite wall, $I = 2W/4\pi d^2$ since the sound is radiated only over ha of a spherical space at distance d. Similarly, for a source at th intersection of two planes at right angles, $I = 4W/4\pi d^2$, and at th intersection of three planes at right angles, $I = 8W/4\pi d^2$. I general, $I = QW/4\pi d^2$, where Q is a factor of 1, 2, 4 or 8, for th cases quoted above, or some appropriate value in the case of source having strongly directional properties. The sound pressur due to the direct field may be found from eqn. (4.62):

$$\frac{p_d^2}{\rho_0 c} = \frac{QW}{4\pi d^2} \tag{4.81}$$

After the first reflection, the sound power remaining,

$$W - W\bar{\alpha} = W(1 - \bar{\alpha}),$$

will contribute to the reverberant field. If the steady energy densit due to this field is homogeneous and has a value of E, the energ absorbed per reflection will be $EV\bar{\alpha}$. If there are n reflections in uni time, the power absorbed will be $EV\bar{\alpha}n$. Now $n = c/L = cS/4V$ where L is the mean free path as given by eqn. (4.78). Under stead state conditions, the rate at which energy is absorbed will be th same as that at which it is supplied, that is

$$W(1 - \bar{\alpha}) = E\bar{\alpha}cS/4$$

From eqn. (4.64),

$$Ec = \frac{p_r^2}{\rho_0 c} = \frac{4W(1 - \bar{\alpha})}{S\bar{\alpha}} \tag{4.82}$$

where p_r is the r.m.s. sound pressure due to the reverberant field Since mean square pressures at a point are generally additiv (eqn. (4.70))

$$\frac{p^2}{\rho_0 c} = \frac{p_d^2}{\rho_0 c} + \frac{p_r^2}{\rho_0 c}$$

$$= \frac{QW}{4\pi d^2} + \frac{4W(1 - \bar{\alpha})}{S\bar{\alpha}} \tag{4.83}$$

and the sound pressure level at the point is given by

$$L_p = 10 \log_{10} \frac{p^2}{(0 \cdot 00002)^2}$$

$$= 10 \log_{10} \frac{\rho_0 c W}{(0 \cdot 00002)^2} \left[\frac{Q}{4\pi d^2} + \frac{4(1-\bar{\alpha})}{S\bar{\alpha}} \right] dB$$

Now $0 \cdot 00002^2/\rho_0 c$ may be taken as having a value of 10^{-12} watt/m^2 for normal air conditions (eqn. (4.63)). Thus, for d in metres and s in metres^2,

$$L_p = 10 \log_{10} \frac{W}{10^{-12}} \left[\frac{Q}{4\pi d^2} + \frac{4(1-\bar{\alpha})}{S\bar{\alpha}} \right] dB$$

$$= L_w + 10 \log_{10} \left[\frac{Q}{4\pi d^2} + \frac{4(1-\bar{\alpha})}{S\bar{\alpha}} \right] dB \qquad (4.84)$$

If the dimensions are in feet, 1 ft^2 may be taken as very nearly 10^{-1} m^2 and

$$L_p = L_w + 10 + 10 \log_{10} \left[\frac{Q}{4\pi d^2} + \frac{4(1-\bar{\alpha})}{S\bar{\alpha}} \right] dB \qquad (4.85)$$

4.11. THE FAN AS A NOISE SOURCE

Whereas sound is any disturbance in an elastic medium capable of being heard, noise is undesired sound. Noise is also defined as a random phenomenon, and certainly the majority of the sound emitted by a fan is random both with respect to frequency and time. It is reasonable to regard the sound from a fan as noise. By far the largest proportion of the noise propagated from the fan inlet or outlet is noise of aerodynamic origin which may have a number of components, such as:

(a) thickness noise due to the passage through the air of blades of finite thickness;

(b) noise due to forces exerted by the fan blades on the air;

(c) rotation noise due to the passage of blades past any fixed point resulting in sound at the blade passage frequency (number of blades × speed of rotation) and harmonic multiples of this frequency;

(d) vortex shedding noise due to flow separation from solid/air boundaries in decelerating flow;

(e) air turbulence noise which is due to shear forces in fluid regions remote from boundaries; and

(f) interference noise due to contact by turbulent wakes on obstructions and guide vanes, the interference by rotating wakes from blades on closely spaced guide vanes also causing discrete frequency components.

Probably the most prominent source of sound in ventilating fans is that due to interaction of fluid forces with solid boundaries mentioned in (d) above. It has been shown[15] that the sound power due to this source may be expressed by

$$W \propto \rho_0 u^6 c^{-3} l^2 f(Re) \qquad (4.86)$$

where l is a typical dimension of the surface concerned. Taking impeller diameter d as a typical dimension for a range of homologous fans, and impeller peripheral velocity, $u = \pi dn$:

$$W \propto \rho_0 n^6 c^{-3} d^8 f(Re)$$

Now $c = (\gamma p_0 / \rho_0)^{\frac{1}{2}}$, so for average ambient air conditions

$$W \propto \rho_0^{5/2} n^6 d^8 f(Re) \qquad (4.87)$$

Introducing the fan laws (eqns. (2.14) and (2.15)), that is, volume flow, $Q \propto nd^3$ and fan pressure, $p \propto \rho_0 n^2 d^2$,

$$W \propto p^{5/2} Q . f(Re) \qquad (4.88)$$

or since air power and fan power $\propto p . Q$, and ignoring $f(Re)$ over a moderate range of variation,

$$W \propto (\text{fan power}) \times (\text{fan pressure})^{3/2} \qquad (4.89)$$

Measurements of the total noise output of fans show that the index of u in eqn. (4.86) is likely to be of the order of 5–5·5 rather than the theoretical value of 6, and this is thought to be due partly to Reynolds number effects.

The total noise emitted by a fan, being almost fully random in character, contains sound energy components covering a very wide range of frequency. When considering noise, only the components in the audible frequency range (from about 20 to about 16,000 c/s) are of interest. Indeed, due to the non-linear response of the human ear to sound pressures of very low and very high frequencies, the frequency range of interest may be regarded as being from about 100 to 10,000 c/s. Since it is difficult to agree on a simple criterion which relates loudness to a single value of sound pressure as read from a measuring instrument, it is customary to make readings for each frequency, or band of frequencies, in the range of interest. The band width considered most practicable for fan noise is the octave band, for which the upper limit of frequency measured is twice the lower limit, the name deriving from the musical pitch interval of an octave. Each octave band is designated by the geometrical mean frequency, $f_{mean} = \sqrt{2}.f_1 = f_2/\sqrt{2}$, f_1 and f_2 being the lower and upper band limits respectively. The standard octave series is based on one band having a mean frequency of 1000 c/s, and the bands of greatest interest so far as fan noise is concerned have mean frequencies of 125, 250, 500, 1000, 2000, 4000 and 8000 c/s. For greater detail, it is common to subdivide octave bands into $\frac{1}{3}$ octave bands, there being three of these for the same frequency range as that of an octave band. The mean frequencies have values in a $3\sqrt{2}$ series (more accurately, a $10\sqrt{10}$ series in practice), every third band having a value coinciding with one in the octave band series. Since there are three $\frac{1}{3}$ octaves to a whole octave, the latter contains the sound energy of three of the narrower bands and the octave band level L_0 may be calculated from eqn. (4.71):

$$L_0 = 10 \log_{10} \left[\text{antilog} \frac{L_A}{10} + \text{antilog} \frac{L_B}{10} + \text{antilog} \frac{L_C}{10} \right]$$

where L_A, L_B and L_C are the $\frac{1}{3}$ octave band levels in the octave. It is not possible to determine the $\frac{1}{3}$ octave band levels from octave band levels. Figure 4.11 shows both octave and $\frac{1}{3}$ octave band spectra for a backward bladed centrifugal fan. It is seen that the $\frac{1}{3}$ octave band spectrum gives the greater definition and appears to show lower general levels. Where the octave band spectrum barely shows the existence of a blade passage frequency, this is quite prominent in the $\frac{1}{3}$ octave spectrum.

So far as it is possible to indicate typical fan spectra, Fig. 4.11 shows comparative spectra for a backward aerofoil bladed centrifugal fan and a contra-rotating axial flow fan for the same duty of 3500 ft^3/min of air at a fan static pressure of 2·5 in. of water. The total sound power level is about the same in each case, but the spectrum shape is rather different. The centrifugal fan has most of its sound energy concentrated at low frequencies, whilst the sound energy of the axial flow fan is spread over a greater frequency range. To the ear, which is less sensitive at low frequencies, the centrifugal fan would probably sound rather quieter than the axial flow fan. It should be pointed out that the accuracy of measurement of this type of noise is of the order of 2 dB owing to random fluctuations in the level from instant to instant, particularly at low frequencies. The blade frequency, too, is more prominent in the noise from certain

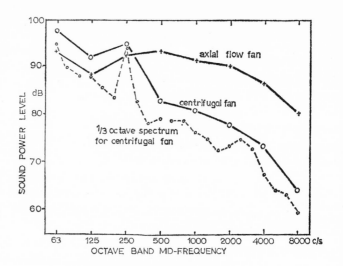

FIG. 4.11. Fan noise spectra.

types of fan, for example, the paddle bladed centrifugal fan; although it may not be particularly evident in an octave band spectrum, it will usually show in a $\frac{1}{3}$ octave band spectrum as can be seen in Fig. 4.11.

The spectra shown in Fig. 4.11 are for sound propagated from the discharge of the fans concerned. The total sound power from the inlet side of each fan will be much the same as from the outlet, but there may be differences of up to about 5 dB in some octave bands. Figure 4.12 shows a comparison between inlet and outlet spectra for an upstream guide vane axial flow fan and a sheet metal backward bladed centrifugal fan. These two fans, it should be noted, do not have the same duty rating, the former having a fan static pressure of about 0·8 in. of water, and the latter of about 2·0 in. of water. The discharge fan sound power seems to be much the same whether the resistance to air flow is on the inlet or on the discharge side of the fan. The same may be said for the inlet fan sound power. Moreover, the form taken by the resistance has little effect unless forming an obstruction in the wake of the direct air stream from fan blades or vanes. There is a change in spectrum level and shape, however,

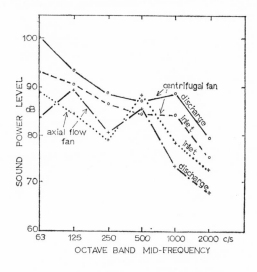

FIG. 4.12. Inlet and outlet noise from fans.

as the fan air duty point changes. Generally, a fan generates least noise at a rating corresponding to maximum fan static efficiency. Also, in general, the quietest fan for a particular air duty is the one with the highest fan static efficiency and not necessarily the largest and slowest running fan as has sometimes been thought. Often a relatively noisy fan, which will probably be inexpensive and compact, with an associated unit silencer may be a suitable choice where noise level is critical.

It seems unlikely that the relationships expressed in eqns. (4.86) to (4.89) will hold in those forms for $\frac{1}{3}$ and whole octave band levels of fan noise. It may be reasonable to assume that as the blade frequency changes, some of the sound energy associated with it will move into another part of the spectrum and, if the speed change is sufficiently great, be evident in another frequency band. The general vortex and turbulence noise appears to change in frequency also but there appears insufficient evidence, at this stage of the art, to draw firm conclusions. A criterion which has been used to relate frequency of sound, f, generated aerodynamically to air velocity v and size of solid body, d, is the Strouhal number, fd/v, which for simple shapes remains constant over fairly wide ranges of frequency. Thus the variation of fan noise frequency with speed may depend approximately on a criterion of this kind.

4.12. NOISE FROM VENTILATING SYSTEMS

The calculation of sound pressure level in a room due to an airconditioning system is clearly of importance. Noise may be generated by the fan, and at points in the system where there is loss of pressure (particularly where this is due to flow separation), and often also at system inlet and discharge grilles. However, there is also some attenuation of fan noise, where this is in excess of locally generated noise, at points such as:

(a) long straight ducts, where sound energy is dissipated in vibration of the sheet metal walls (Table 4.1),

TABLE 4.1. Attenuation of Sound in Sheet Metal Ducts at Air Velocities of less than 2000 ft/min (11 m/s)[16]

Duct size (in.)	Octave band mid-frequency (c/s)						
	125	250	500	1000	2000	4000	8000
12×12	0·26	0·20	0·16	0·08	0·12	0·12	0·16 dB/ft
12×24	0·36	0·16	0·06	0·12	0·12	0·12	0·16 dB/ft

Chaddock, J. B., *Refrigerating Engineering*, **67**, 37 (1959).

(b) at bends, where multiple reflection and absorption may occur (Table 4.2),

(c) at the entrance to branch ducts, since the total sound power is distributed to the branches. The attenuation at an individual branch may be expressed approximately as:

$$\text{Attenuation} = 10 \log_{10} \frac{\text{total duct cross-section area}}{\text{branch duct cross-section area}} \text{ dB} \qquad (4.90)$$

TABLE 4.2. Approximate Attenuation by Round or Square Bends fitted with Turning Vanes (Air Velocity not exceeding 1500 ft/min)[17]

Duct size (in.)	Octave band mid-frequency (c/s)						
	125	250	500	1000	2000	4000	8000
5 to 10	0	0	0	1	2	3	3 dB
11 to 20	0	0	1	2	3	3	3
21 to 40	0	1	2	3	3	3	3
41 to 80	1	2	3	3	3	3	3

At the open end of a duct there is further effective attenuation of sound since some reflection of the sound energy back along the duct takes place. The amount of energy reflected depends on the re-

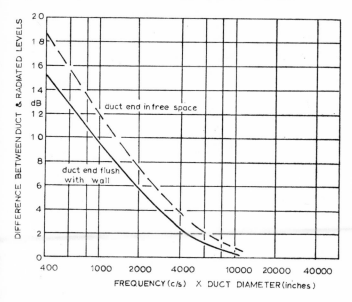

FIG. 4.13. Effective attenuation due to reflection of sound at the open end of a pipe.

lationship between the size (and to some extent the shape) of the discharge or inlet orifice and the frequency of the sound. The order of magnitude of this effect is given in Fig. 4.13.

The sound power radiated into a room from the open end of a
duct is found by subtracting from the fan sound power level (at the
fan inlet or outlet, whichever is relevant) the total attenuation due to
straight ducts, bends, branches and end reflection. Very often there
is a decorative grille at the end of the duct which will itself generate
noise, the power of which must be added to that calculated above by

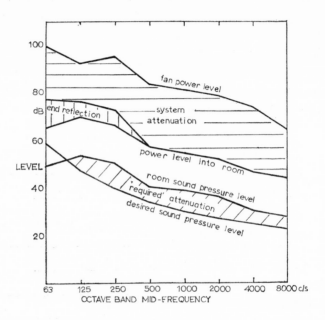

FIG. 4.14. Transmission of sound in an air-conditioning system.

using eqn. (4.71). The sound pressure level at any point within the
room may now be calculated by using eqn. (4.85). If there are n
outlets of equal sound power, eqn. (4.82) may be rewritten

$$Ec = \frac{p_r^2}{\rho_0 c} = \frac{4nW(1-\bar{\alpha})}{S\bar{\alpha}} \qquad (4.91)$$

The direct field from any outlet other than the nearest will generally
be negligible and equation may be written

$$L_p = L_w + 10 + 10\log_{10}\left[\frac{Q}{4\pi d^2} + \frac{4n(1-\bar{\alpha})}{S\bar{\alpha}}\right] \text{dB}, \qquad (4.92)$$

where L_w is the sound power level at any one outlet. An example of this process is shown diagrammatically applied to the spectrum of fan noise in Fig. 4.14, calculations being performed for each octave band.

4.13. ATTENUATION OF SOUND BY LINED DUCTS

Having performed the calculations outlined in the previous section, a decision on the acceptability of the resulting sound pressure level

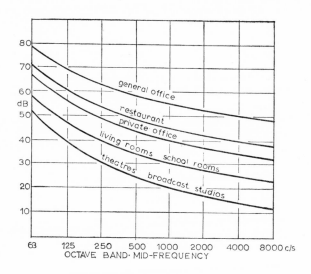

FIG. 4.15. Suggested noise ratings from some types of environment.

must be reached. This will depend on the nature of the sound spectrum and the use of the space being conditioned. The type of noise from air moving systems will often be acceptable if the spectrum conforms to the shape of one of those given in Fig. 4.15 for a few common environments. If the selected spectrum is exceeded by more than the odd decibel it is desirable to provide additional attenuation in the system. This may take the form of a unit " silencer " (often fitted close to the fan) where substantial attenuation is required. Where only a moderate degree of reduction is required, this can often be achieved by the use of absorptive duct linings. These may satis-factorily be designed using the expression

$$\text{attenuation} = \frac{S\alpha^{1\cdot4}}{A}\,\text{dB} \qquad (4.93)$$

where S is the total exposed surface area of absorptive material whose coefficient of absorption is α, and A is the cross-sectional area of the airway. Some values of absorption coefficients for suitable lining materials are given in Table 4.3.

TABLE 4.3. Absorption Coefficients of Some Materials Suitable for Duct Linings

Material	Frequency in cycles per second					
	125	250	500	1000	2000	4000
1 in. sprayed asbestos		0·60	0·65	0·60	0·60	
2 in. sprayed asbestos		0·85	0·94	0·90	0·80	
2 in. mineral wool, 3 lb/ft³	0·25	0·55	0·70	0·75	0·80	0·90
2 in. mineral wool, 5 lb/ft³	0·25	0·65	0·80	0·85	0·90	0·90
1 in. glass fibre, bonded	0·10	0·35	0·55	0·75	0·80	0·50
2 in. glass fibre, bonded	0·20	0·50	0·75	0·85	0·70	0·65

The dimensions refer to the thickness of the material.

4.14. EXAMPLES

1. A fan running at 700 rev/min is driven by an electric motor running at 1470 rev/min. The assembly is mounted on steel spring isolators which deflect by 0·5 in. Find the transmissibility to vibration forces resulting from out of balance of each.

Natural frequency of isolators, $f_0 = 3\cdot13\sqrt{(1/\delta)} = 3\cdot13\sqrt{(1/0\cdot5)}$
$\qquad\qquad = 4\cdot43$ c/s (eqn. (4.39)).

Frequency of forced vibration due to rotational out of balance of the fan impeller $= 700/60 = 11\cdot67$ c/s.

Transmissibility $= \dfrac{1}{1-f^2/f_0^2} = \dfrac{1}{1-(11\cdot67/4\cdot43)^2} = \dfrac{1}{1-6\cdot95}$
$\qquad\qquad = -0\cdot168$ (see eqn. (4.15)).

Frequency of forced vibration due to rotational out of balance of motor rotor $= 1470/60 = 24\cdot5$ c/s.

Transmissibility $= \dfrac{1}{1-(24\cdot5/4\cdot43)^2} = \dfrac{1}{1-30\cdot5} = -0\cdot0339$.

The negative sign indicates a phase difference of 180° between impressed and transmitted forces.

2. A 50 c/s a.c. transformer, weighing 3 tons, is to be mounted on cork pads 3 in. thick, the cork having a dynamic modulus of elasticity of 1500 lb/in² at loadings in the range 3–8 lb/in². Calculate the area of cork necessary to limit transmission of magnetic vibration forces to 10%.

Observing the negative sign for a value of transmissibility of less than unity:

$$-0{\cdot}1 = 1/(1-f^2/f_0^2);$$

$$f^2/f_0^2 = 1+1/0{\cdot}1 = 11;$$

$$f/f_0 = \sqrt{11} = 3{\cdot}32.$$

Frequency of forced vibration = $2 \times 50 = 100$ c/s.

Frequency of natural vibration = $100/3{\cdot}32 = 30{\cdot}2$ c/s.

Required deflection of mounting = $(3{\cdot}13/f_0)^2$ (eqn. (4.39))

$$= (3{\cdot}13/30{\cdot}2)^2 = 1/93 \text{ in.}$$

Area of cork required $= \dfrac{\text{load} \times \text{thickness}}{\text{modulus} \times \text{deflection}}$

$$= \frac{3 \times 2240 \times 3}{1500 \times 1/93} = 1250 \text{ in}^2 \text{ or } 8{\cdot}7 \text{ ft}^2.$$

3. A fan weighing 500 lb is driven by a two-speed motor running at either 460 or 975 rev/min. The amplitude of vibration at the higher speed is 0·0110 in., and at the lower speed is 0·0136 in. Assuming that the isolators are not fully compressed and are undamped, calculate the extra weight to be added to give acceptable isolation at the lower speed.

Vibrational force due to rotational out of balance is proportional to (rotational speed)² (eqn. (4.1)).
Using eqn. (4.13),

$$-0{\cdot}0136 = \frac{F}{k} \cdot \frac{1}{(1-f_1^2/f_0^2)} = \frac{F}{k} \cdot \frac{f_0^2}{(f_0^2-f_1^2)}$$

$$-0{\cdot}0110 = \frac{F}{k} \frac{f_2^2}{f_1^2} \cdot \frac{1}{(1-f_2^2/f_0^2)} = \frac{F}{k} \frac{f_2^2}{f_1^2} \cdot \frac{f_0^2}{(f_0^2-f_2^2)}$$

$$\frac{0{\cdot}0136}{0{\cdot}0110} \cdot \frac{(1460)^2}{(975)^2} = 2{\cdot}78 = \frac{f_2^2-f_0^2}{f_1^2-f_0^2}$$

$$1{\cdot}78 f_0^2 = (2{\cdot}78-f_2^2/f_1^2)f_1^2 = (2{\cdot}78-2{\cdot}24)f_1^2 = 0{\cdot}54 f_1^2$$

$$f_0 = 0{\cdot}55 f_1 = 0{\cdot}55 \times 975 = 537 \text{ c/min.}$$

A reasonable value for f_0 would be (section 4.2) 975/2·5, or 390 c/min. From eqn. (4.11), $f_0 \propto 1/\sqrt{W}$ from which the isolator loading should be

$$500 \times (537/390)^2 = 950 \text{ lb.}$$

The extra load required = $950 - 500 = 450$ lb.

4. A load of 100 lb is placed on four pieces of cork, each a 2 inch cube, and resonance is found at 2200 c/min when the amplitude of vibration is 0·010 in. The maximum value of the impressed force is 30 lb. Find the dynamic modulus of elasticity and the damping ratio, c/c_e for the cork.

Resonance occurs when the frequency of forced vibration is the same as the frequency of natural vibration, which in this case is 2200 c/min. The cork deflection corresponding to this will be $(3·13 \times 60/2200)^2 = 0·0073$ in. (Note that this may not be the same as the observed static deflection.)

$$\text{Dynamic modulus} = \frac{\text{stress}}{\text{strain}} = \frac{100 \times 2}{4 \times 2 \times 2 \times 0·0073} = 1715 \text{ lb/in}^2.$$

Amplitude of vibration at resonance may be found from eqn. (4.25) by putting $\omega/\omega_0 = 1(f/f_0 = 1)$ as, $a = \dfrac{F}{2k} \cdot \dfrac{c_e}{c}$

$$c/c_e = F/2ak$$

Stiffness of cork, k = load/area = $100/0·0073 = 13,700$ lb/in.

$$c/c_e = 30/(2 \times 0·010 \times 13,700) = 0·109.$$

5. The background sound pressure level at a point in a room is measured and found to be 56 dB. The sound from a ventilating fan increases this to 58 dB. What would be the sound pressure level due to the fan alone?

From eqn. (4.72),

$$A = 10 \log_{10} \left(\log^{-1} \frac{C}{10} - \log^{-1} \frac{B}{10} \right) \text{ dB,}$$

where A is the level due to the fan alone, C is the level due to fan and background, B is the level due to background alone.

$A = 10 \log_{10}(\log^{-1}58/10 - \log^{-1}56/10) = 10 \log_{10}(6·31 \times 10^5 - 3·98 \times 10^5)$

$= 10 \log_{10}(2·33 \times 10^5) = 10 \times 5·37 = 53·7$ dB.

In practice, fractions of a decibel have no real significance since measurements can rarely be made to better than the nearest whole decibel.

6. In an anechoic chamber, the mean sound pressure level at a distance of 8 ft from the centre of a fan is 73 dB. Find the sound power output and the sound power level of the fan. Estimate the reduction in sound power level of the fan if the fan speed is reduced by 10%.

In an anechoic chamber there is little or no reflected sound and free field conditions exist. Sound pressure level is numerically equal to sound intensity level and from this, sound intensity (which is sound power per unit area) may be found from eqn. (4.66), as $I = 10^{-12} \log^{-1} 73/10$
$$= 10^{-12} \times 10^7 \times 2 = 2 \times 10^{-5} \text{ watt/m}^2.$$

Since 1 m = 3·28 ft, 8 ft = 8/3·28 m, and sound power, $W = I \times 4\pi d^2 = 2 \times 10^{-5} \times 4\pi \times (8/3·28)^2 = 1·5 \times 10^{-3}$ watt.

Sound power level $= 10 \log_{10} (1 \cdot 5 \times 10^{-3}/10^{-12})$ dB (eqn. (4.65))

$\qquad\qquad\qquad = 10 \times 9 \cdot 18 = 91 \cdot 8$ dB $\simeq 92$ dB.

Reference to section 4.11 and eqn. (4.87) indicates that sound power from a fan varies as (tip speed)^6 in theory. It is found by experiment that the exponent of tip speed is more nearly 5. Thus $W_2/W_1 = (\pi n_2 d)^5/(\pi n_1 d)^5 = (n_2/n_1)^5$.

Change in power level with change in speed will be

$10 \log_{10} (W_2/10^{-12}) - 10 \log_{10} (W_1/10^{-12}) = 10 \log_{10} (W_2/W_1)$

$\qquad\qquad\qquad\qquad\qquad\qquad = 10 \log_{10} (n_2/n_1)^5 = 50 \log_{10} (n_2/n_1),$

and substituting the value of 0·9 for n_2/n_1, the change in level will be

$\qquad 50 \log_{10} (0 \cdot 9) = 50 \times \bar{1} \cdot 954 = -50 + 47 \cdot 7 = -2 \cdot 3$ dB,

that is, there will be a reduction in sound power level of 2·3 dB.

7. A fan discharges into a fully reverberant room whose volume is 10,000 ft^3 and surface area is 2700 ft^2. The reverberation time of the room at 1000 c/s is 5 sec and the sound pressure level throughout the room due to the fan in the 1000 c/s octave band is 95 dB when the air temperature is 70°F. Find the sound power level of the fan in the 1000 c/s octave band.

From the reverberation time, the average absorption coefficient $\bar{\alpha}$ may be found (eqn. (4.80)). Since $\bar{\alpha}$ has a low value in a reverberant room,

$$\bar{\alpha} = \frac{55 \cdot 3 V}{cSt} .$$

From eqn. (4.53), velocity of sound in air at 70°F will be

$$c = 49 \cdot 1 \sqrt{(460 + 70)} = 1130 \text{ ft/s},$$

$$\bar{\alpha} = \frac{55 \cdot 3 \times 10,000}{1130 \times 2700 \times 5} = 0 \cdot 0363.$$

And from eqn. (4.85),

$$L_w = L_p - 10 - 10 \log_{10} \left[\frac{Q}{4 \pi d^2} + \frac{4(1 - \bar{\alpha})}{S \bar{\alpha}} \right] \text{ dB}.$$

In a fully reverberant room where the sound pressure level is uniform, the direct field term in the above equation is negligible, thus

$$L_w = L_p - 10 - 10 \log_{10} (4(1 - \bar{\alpha})/S \bar{\alpha}) \text{ dB}$$

$$= 95 - 10 + 10 \log_{10} (2700 \times 0 \cdot 0363/3 \cdot 96)$$

$$= 85 + 10 \log_{10} 24 \cdot 5 = 85 + 10 \times 1 \cdot 39$$

$$= 85 + 14 = 99 \text{ dB}.$$

8. The term specific sound power level may be applied to a fan in the same sense as specific speed. A fan having a duty of 20,000 ft^3/min at a pressure of 3·5 in. of water has a sound power level of 110 dB. Assuming that sound power varies as (tip speed)^5, find the specific sound power level for the homologous series to which the fan belongs.

From eqns. (4.86) to (4.88), it can be seen that if sound power varies as (tip speed)^6, sound power W varies as $p^{5/2}.Q.f(Re)$. By the same arguments, if

$W \propto u^5$, then since $p \propto u^2$ and $Q \propto n$, $W \propto p^2 Q$. Over a moderate range of volume and pressure, the $f(Re)$ term may be neglected. Taking specific sound power W_s to be the sound power of a fan in the same homologous series which gives a duty of 1 ft³/min at a pressure of 1 in water;

$$\frac{W_s}{W} = \frac{1}{p^2 Q}, \quad \text{or} \quad W_s = W/p^2 Q.$$

Specific sound power level $= 10 \log_{10}(W_s/10^{-12})$

$$= 10 \log_{10}(W/10^{-12}) - 10 \log_{10}(p^2 Q) \text{ dB}$$

$$= L_w - 20 \log_{10}p - 10 \log_{10}Q \text{ dB},$$

and for the series of fans given

$$L_s = 110 - 20 \log_{10}(3 \cdot 5) - 10 \log_{10}(20,000)$$

$$= 110 - 20 \times 0 \cdot 544 - 10 \times 4 \cdot 301$$

$$= 110 - 11 - 43 = 56 \text{ dB}.$$

9. The sound power level of a fan in the 500 c/s octave band is 96 dB. It supplies air to a system comprising a straight duct of 12 in. by 24 in. cross-section and 100 ft long, a junction to 50 ft of 12 in. by 12 in. duct having three bends whose total attenuation is 3 dB, and finally to an outlet in the ceiling of a room having a diffuser which generates a sound power level in the 500 c/s octave band of 48 dB at the design air velocity. The room is 30 ft by 20 ft by 10 ft high and has surfaces having an average absorption coefficient of 0·08. Find the sound pressure level in the room at a point 8 ft from the ceiling diffuser in the 500 c/s octave band.

All calculations are carried out for the 500 c/s octave band.
Attenuation in 24 in. × 12 in. duct = $100 \times 0 \cdot 06 = 6$ dB (Table 4.1).
Attenuation at junction = $10 \log_{10}[24 \times 12/(12 \times 12)]$

$$= 10 \log_{10}(2) = 3 \text{ dB (eqn. (4.90))}.$$

Attenuation in 12 in. × 12 in. duct = $50 \times 0 \cdot 16 = 8$ dB.
Power level due to bends = 3 dB.
End reflection effect from Fig. 4.13

$$\left(f \times d = 500 \times \frac{4}{\pi} \times 12 = 7650 \text{ in/s} \right)$$

is seen to be about 1 dB.
Power level at outlet due to fan = $96 - (6+3+8+3+1)$

$$= 96 - 21 = 75 \text{ dB}.$$

Combined power level due to fan and diffuser (eqn. (4.71)):

$$= 10 \log_{10}(\log^{-1}7 \cdot 5 + \log^{-1}4 \cdot 8) = 10 \log_{10}(3 \cdot 16 \times 10^7 + 6 \cdot 31 \times 10^4)$$
$$= 75 \text{ dB}.$$

It is seen that the sound power generated by the diffuser has negligible effect in this example.

Sound pressure level in the room:

$$L_p = L_w + 10 + 10 \log_{10}\left(\frac{Q}{4\pi d^2} + \frac{4(1-\bar{\alpha})}{S\bar{\alpha}}\right) \text{ dB.}$$

If the outlet is in the ceiling (plane surface),

$$Q = 2 \quad \text{and} \quad Q/4\pi d^2 = 2/(4\pi \times 8^2) = 0.0025.$$

Room surface area $= 2 \times 30 \times 20 + 2 \times 30 \times 10 + 2 \times 20 \times 10$
$$= 1200 + 600 + 400 = 2200 \text{ ft}^2$$

and $4(1-\bar{\alpha})/S\bar{\alpha} = 4 \times 0.92/(2200 \times 0.08) = 0.0209.$

Thus $L_p = 75 + 10 + 10 \log_{10}(0.0025 + 0.0209)$

$$= 85 + 10 \log_{10}(0.0234) = 85 + 10(-2 + 0.37)$$

$$= 65 + 3.7 = 68.7 \quad \text{or} \quad 69 \text{ dB.}$$

10. Find the length of 12 in. by 24 in. duct to be lined on the inside with material, 1 in. thick and having an absorption coefficient of 0.55, to give an attenuation of 10 dB.

Area of lining $= 2 \times [(12-2) + (24-2)] \times L = 64\,L \text{ in}^2.$

Area of airway $= (12-2) \times (24-2) = 220 \text{ in}^2.$

Then $10 = \dfrac{64\,L \times 0.55^{1\cdot4}}{220} = \dfrac{64\,L \times 0.433}{220} = 0.126\,L.$

Length of lining $L = 10/0.126 = 79.5$ in.

CHAPTER 5

FAN DESIGN

5.1. WORK DONE BY A ROTATING IMPELLER

Consider a rotating impeller to which air approaches at radius r_1 with some absolute velocity v_1, and enters with a relative velocity w_1 and at angle β_1 to the impeller periphery. A vane is fixed to the impeller at an angle β_1 at the point of entry, allowing the air to pass

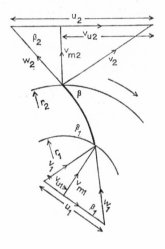

FIG. 5.1. Blade velocity triangles.

over the surface at this point without change of direction. The air passes along the vane, leaving it with a relative velocity w_2 at the impeller outer radius r_2, where the angle made by the vane to the periphery is β_2. The absolute leaving velocity of the air (relative to some fixed point outside the impeller), v_2, will depend on w_2 and the impeller peripheral velocity $u_2 = \omega r_2$. Considering the velocities to be coplanar, velocity triangles may be constructed to show vectorially

122

the relative velocities at inlet and outlet of the impeller, as shown in Fig. 5.1. The tangential components of absolute velocities v_1 and v_2 are v_{u1} and v_{u2}, whilst the radial (or meridional) components are v_{m1} and v_{m2}. The work done on the air by the impeller will be the difference between the energy in the air leaving and the energy in the air entering the impeller in the direction of rotation.

Energy in air leaving the impeller = torque × angular displacement

= rate of change of (tangential momentum × radius × angular displacement)

= tangential momentum × radius × rate of change of angular displacement

$$= \frac{W}{g} \cdot v_{u2} \times r_2 \times \omega_2 \text{ for a flow of weight of air } W.$$

Similarly, energy in air entering the impeller

$$= \frac{W}{g} \cdot v_{u1} \times r_1 \times \omega_1$$

Thus work done by the impeller

$$= \frac{W}{g} (v_{u2} u_2 - v_{u1} u_1)$$

The head developed by the impeller may be defined as the height to which the same weight of air may be raised by the same amount of work, that is

$$WH = \frac{W}{g} (v_{u2} u_2 - v_{u1} u_1)$$

or

$$H = \frac{v_{u2} u_2}{g} - \frac{v_{u1} u_1}{g} \tag{5.1}$$

Equation (5.1) is sometimes referred to as Euler's equation for an impeller, and the head H as Euler's head.

Using the relationship between head and pressure from eqn. (1.6),

$$p = wH = \frac{w}{g} v_{u2} u_2 - \frac{w}{g} v_{u1} u_1$$

$$= \rho v_{u2} u_2 - \rho v_{u1} u_1 \tag{5.2}$$

The expression for head given by eqn. (5.1) may also be expressed in terms of other velocities, for

$$v_m^2 = v^2 - v_u^2 = w^2 - (u - v_u)^2 = w^2 - u^2 + 2uv_u - v_u^2$$

or

$$v_u u = \frac{u^2 + v^2 - w^2}{2}$$

and

$$H = \frac{v_{u2} u_2}{g} - \frac{v_{u1} u_1}{g}$$

$$= \frac{u_2^2 - u_1^2}{2g} + \frac{v_2^2 - v_1^2}{2g} + \frac{w_1^2 - w_2^2}{2g} \qquad (5.3)$$

Analysing this result into the three components:

$\dfrac{u_2^2 - u_1^2}{2g}$ is the increase of head due to the forced vortex caused by the impeller rotation (eqn. (1.57)),

$\dfrac{v_2^2 - v_1^2}{2g}$ is the increase in velocity head of the air in passing through the impeller, and

$\dfrac{w_1^2 - w_2^2}{2g}$ is the regain of static head consequent on reduction in relative velocity in the air passing through the impeller.

So far, it has been assumed that air can enter the impeller inlet in any direction. Since the vast majority of centrifugal fans do not have inlet guide vanes, air will tend to enter in a radial direction as uniform pressure distributions might be expected at both inlet and outlet peripheries. In such cases $v_{u1} = 0$ and the total pressure developed

$$p = \rho u_2 v_{u2} \qquad (5.4)$$

Variable guide vanes are sometimes used for regulation of volume flow. These reduce the amount of work done on the air by the impeller by imparting rotation to the air in the direction of the impeller rotation, thus giving finite values to v_{u1}. It is to be expected that the work done by an impeller by having air entering with pre-rotation in an opposite direction to that of the impeller rotation (and giving negative values of v_{u1}) will be increased. Axial flow fans with

upstream guide vanes may utilize this principle but it is rarely used
for centrifugal fans.

5.2. CENTRIFUGAL FAN VELOCITY TRIANGLES

For normal centrifugal fans, it is seen from eqn. (5.4) that the work
done by an impeller increases as v_{u2} increases. From Fig. 5.2, it can
be seen that as the blade angle β_2 increases the ratio v_{u2}/u_2 also
increases. For the type of impeller shown:

(i) backward curved; $\beta_2 < 90°$, $v_{u2} < u_2$.

(ii) radial tipped; $\beta_2 = 90°$, $v_{u2} = u_2$.

(iii) forward curved; $\beta_2 > 90°$, $v_{u2} > u_2$.

Clearly, the greatest pressure development for a given peripheral
velocity u_2 is to be expected from a forward curved impeller.

The volume flow through the impeller will be the product of
velocity and the area normal to the direction of flow. Taking the
radial velocity v_m,

Volume flow,

$$Q = \pi d_2 b_2 v_{m2} = \pi d_1 b_1 v_{m1} \tag{5.5}$$

where b_2 and b_1 are the blade widths (axially) at diameters d_2 and d_1
respectively. In practice, v_{m1} and v_{m2} are very nearly the same in
value, and of the order of $0·2u_2$ (Table 5.1). Ideally, the velocity
through the impeller inlet (or " eye ", as it is sometimes called), v_0,
should not be more than v_{m1}. However, this cannot often be achieved
since the inlet diameter becomes too large, and v_0 is generally of the
order of twice v_{m1}, or $0·4u_2$.

Since volume flow is a function of v_{m2}, theoretical characteristics
of impeller total pressure and power requirements may be derived,
assuming radial entry.

(a) $\beta_2 < 90°$. Figure 5.2 shows that

$$u_2 - v_{u2} = v_{m2} \cot \beta_2$$

or

$$v_{u2} = u_2 - v_{m2} \cot \beta_2$$

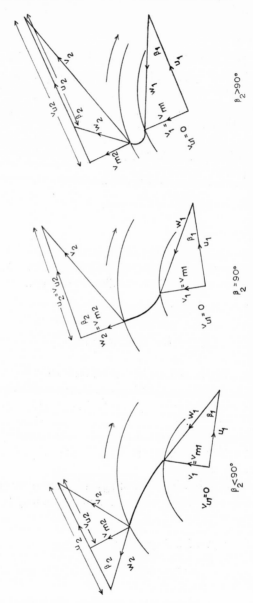

FIG. 5.2. Velocity triangles for centrifugal impeller blade forms.

mpeller total pressure, $p = \rho u_2 v_{u2} = \rho u_2^2 - \rho u_2 v_{m2} \cot \beta_2.$

Volume flow, $\qquad Q = \pi d_2 b_2 v_{m2}.$

Thus

$$p = \rho u_2^2 - \rho u_2 \frac{Q}{\pi d_2 b_2} \cot \beta_2 \qquad (5.6)$$

This may also be written in terms of non-dimensional performance coefficients (eqns. (2.20) to (2.23))

$$\psi = \frac{p}{\frac{1}{2}\rho u_2^2}, \qquad \phi = \frac{4Q}{\pi d_2^2 u_2}$$

and substituting these values in eqn. (5.6),

$$\psi = 2\left(1 - \phi \frac{d_2}{4b_2} \cot \beta_2\right) \qquad (5.7)$$

Power coefficient

$$\lambda = \phi\psi = 2\phi\left(1 - \frac{\phi d_2}{4b_2} \cot \beta_2\right) \qquad (5.8)$$

(b) $\beta_2 = 90°.$

$$u_2 = v_{u2}$$

$$p = \rho u_2 v_{u2} = \rho u_2^2 \qquad (5.9)$$

and

$$\psi = 2 \qquad (5.10)$$

$$\lambda = 2\phi \qquad (5.11)$$

(c) $\beta_2 > 90°.$

$$v_{u2} - u_2 = v_{m2} \cot(180° - \beta_2)$$

$$v_{u2} = u_2 + v_{m2} \cot(180° - \beta_2)$$

from which

$$p = \rho u_2 v_{u2} = \rho u_2^2 + \rho u_2 v_{m2} \cot(180° - \beta_2)$$

$$= \rho u_2^2 + \frac{\rho u_2 Q}{\pi d_2 b_2} \cot(180° - \beta_2) \qquad (5.12)$$

and

$$\psi = 2\left[1 + \frac{\phi d_2}{4b_2} \cot(180° - \beta_2)\right] \qquad (5.13)$$

$$\lambda = 2\phi\left[1 + \frac{\phi d_2}{4b_2} \cot(180° - \beta_2)\right] \qquad (5.14)$$

These theoretical characteristics are plotted in Fig. 5.3 and are seen to differ in shape from those shown in Fig. 2.3 for actual fans. This is due to the fact that there are losses in the fan which are of square law form and vary as ϕ^2. Such losses tend to bend the ψ/ϕ curves in a downward direction, and the power curves in an upward direction. Also, the curves of Fig. 5.3 are for the impeller only and not a complete fan. It is found on test, too, that a forward curved

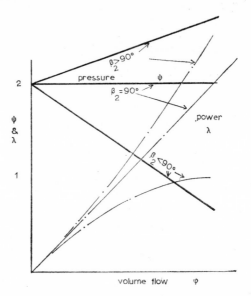

Fig. 5.3. Theoretical characteristics of centrifugal impellers.

fan gives a higher value of ψ than 2 when $\phi = 0$, a fact which is not easy to explain.

It is seen from eqn. (5.7) that when $\phi d_2 \cot \beta_2 / 4b_2 = 1$, the pressure developed is negative, implying that the fan has become a turbine.

5.3. EFFECT OF A FINITE NUMBER OF BLADES

The simple expressions so far derived assume the air to follow the blade profile exactly. This can only be justified if the number of blades is infinite. One cause of deviation of flow from the ideal is interblade circulation. If a container of fluid is whirled with an ngular velocity of ω at the end of an arm, as shown in Fig. 5.4, a article of the fluid such as A tends to remain stationary. However, elative to the container it appears to be rotating in a direction opposite to that of the arm, that is, with an angular velocity of $-\omega$. At any radius a, within the container, it has a relative velocity of $-a\omega$. It has been suggested[18] that the air between adjacent blades f a radial flow impeller behaves in a somewhat similar manner, and hat this motion is superimposed on the outward flow between the blades. The effect of this is to reduce the value of v_{u2} to v'_{u2}, where

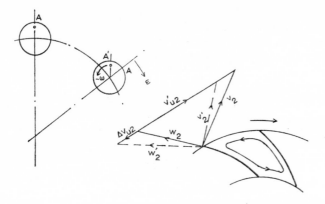

FIG. 5.4. Interblade circulation.

$_{u2}' = v_{u2} - a\omega$. The effective radius of the area between blades is difficult to assess, but is generally taken as half of the perpendicular distance between the blade tangents at the impeller periphery, that is

$$a = \frac{\pi d_2 \sin \beta_2}{2z} \tag{5.15}$$

where z is the number of blades.

Since $\omega = u_2/r_2 = 2u_2/d_2$

$$a\omega = \frac{\pi d_2 \sin \beta_2}{2z} \cdot \frac{2u_2}{d_2} = \frac{\pi \sin \beta_2}{z} \cdot u_2$$

and

$$v'_{u2} = v_{u2} - \frac{\pi \sin \beta_2}{z} \cdot u_2 \tag{5.16}$$

The effect of interblade circulation is to reduce the amount of work, and hence the pressure developed, by a fan impeller. Other suggested corrections to v_{u2} in place of $a\omega$ are[19]

$$\Delta v_{u2} = \frac{0 \cdot 72 z u_2}{\pi r^2}$$

$$\Delta v_{u2} = \left(k' - \frac{\cot \beta_2}{u_2} . v_{m2} \right) u_2$$

where k' is a constant for a particular design.

Here Δv_{u2} is defined by $v'_{u2} = v_{u2} - \Delta v_{u2}$, where v_{u2} has the theoretical value. Impeller total pressure,

$$p = \rho u_2 v'_{u2} \tag{5.17}$$

It should be noted also that the blade passage cross-sectional area will be

$$\pi db - ztb/\sin \beta \tag{5.18}$$

where t is the thickness of each blade. In small fans thick blades may cause constriction, particularly at the blade inlet.

5.4. CENTRIFUGAL FAN CASINGS

Air leaves the impeller of a centrifugal fan with both radial and tangential components of velocity. The absolute velocity is often undesirably high. Whilst in a system of ducts static pressure may be converted to velocity pressure with little loss, the converse is not necessarily true, and it is desirable to have as high a fan static pressure as possible. A good casing must reduce the velocity of the air leaving the impeller, v_2, to the casing outlet velocity, v_3, with as little loss as possible, whilst permitting convenient connection to any duct on the discharge.

In section 1.9 it has been shown that a free vortex for which $v_u r$ is constant, permits flow without loss and ideally a casing should be designed to permit such flow. The casing profile will be a streamline of a free vortex. Referring to Fig. 5.5, at any radius r in the casing

$r = v_{u2}r_2 =$ constant, from which $r_2/r = v_u/v_{u2}$. For a casing
width of b, the volume flow,

$$Q = 2\pi r b v_m = 2\pi r_2 b_2 v_{m2}$$

$$\frac{v_m}{v_{m2}} = \frac{r_2 b_2}{rb} = \frac{b_2 v_u}{b v_{u2}} \tag{5.19}$$

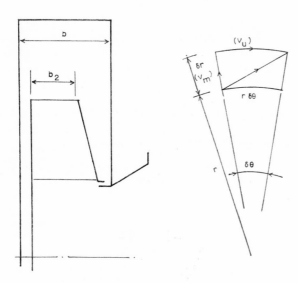

FIG. 5.5. Flow in centrifugal fan casings.

In order to maintain the casing profile to that of a streamline,
$[(r+\delta r)\delta\theta]/\delta r = (v_u/v_m)$, and in the limit, and ignoring small
quantities of the second order,

$$\frac{dr}{r} = \frac{v_m}{v_u}.d\theta = \frac{b_2 v_{m2}}{b v_{u2}}.d\theta$$

(from eqn. (5.19)) and, on integration,

$$\log_e\frac{r}{r_2} = \frac{b_2 v_{m2}}{b v_{u2}}(\theta_2-\theta_1) \tag{5.20}$$

This may be expressed in terms of pressure coefficient ψ, sinc

$$v_{m2}/v_{u2} = (u_2 - v_{u2}) \tan \beta_2 / v_{u2} = (u_2/v_{u2} - 1) \tan \beta_2 \quad (\beta_2 < 90°)$$

$$= (2/\psi - 1) \tan \beta_2$$

Thus

$$\log_e \frac{r}{r_2} = \frac{b_2}{b}(2/\psi - 1) \tan \beta_2 (\theta_2 - \theta_1) \tag{5.21}$$

The ratio b/b_2 is of the order of 2·5 for backward curved fans, an
about 1·25 for forward curved fans. The loss due to a sudden expan
sion from b_2 to b seems to be less than would be expected and th
extra width reduces the casing radius to a minimum. Even so, a fre
vortex casing is often uneconomically large and a further moderat
sacrifice in efficiency is often accepted to reduce the casing stil
further. Frequently, an arithmetical spiral of the form $r = r_2(1 + k\theta$
is used, where, if θ is in degrees, k is of the order of 0·0023 for back
ward curved fans and 0·0020 for forward curved fans. Anothe
simple construction often used is shown in Fig. 5.6, consisting o
four circular arcs. It is not essential that the offsets a be the same i
magnitude.

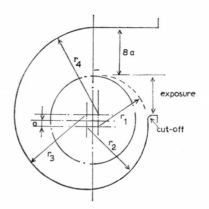

Fig. 5.6. Geometrical construction for casing contour.

Keeping the overall dimensions to a minimum results in a larg
impeller "exposure" (Fig. 5.6) of up to half the impeller diametei

uch large exposures tend to permit reverse flow in the fan outlet at
he cut-off point, and many manufacturers fit a throat plate to
educe vortex formation at this point whilst having little effect on
ir velocity at the outer edge of the casing. But it must be noted that
he air velocity distribution at the outlet of an average centrifugal
an is far from uniform. The radial clearance at the cut-off is of the
rder of 10% of the impeller diameter. Values of less than this tend to
roduce increased noise at blade passage frequency (number of
lades × revolutions per second) and may also reduce fan
erformance.

5.5. LOSSES IN CENTRIFUGAL FANS

The actual performance of a centrifugal fan at the design point of
peration differs from that predicted by Euler's equation, eqn. (5.1).
art of this difference can be accounted for by an adjustment for
nterblade circulation (section 5.3) which results in a reduction of the
work done by the impeller. Amongst other factors contributing to a
eduction in output are:

(a) Internal volumetric leakage between the impeller inlet and
asing inlet, and also where the drive shaft enters the casing. Inlet
eakage, which is likely to be the more serious, may be considered in
imilar fashion to flow through an orifice, that is, leakage volume
$Q_L = C_d A \sqrt{(2p_s/\rho)}$, where C_d is a discharge coefficient, and p_s is
he static difference across the clearance between impeller and casing,
. This may be written

$$Q_L = C_d \pi d_1 \delta \sqrt{\left(\frac{2p_s}{\rho}\right)} \qquad (5.22)$$

If the edges of the gap are sharp, it is reasonable to suppose that
he value of C_d will be of the order of 0·6. In terms of performance
oefficients,

$$\phi_L . \frac{\pi d_2^2}{4} . u_2 = C_d \pi d_1 \delta \sqrt{\left(\frac{2\psi_s' . \frac{1}{2}\rho u_2^2}{\rho}\right)}$$

$$\phi_L = 4C_d . \frac{d_1^2}{d_2^2} . \frac{\delta}{d_1} \sqrt{\psi_s'} \qquad (5.23)$$

The static pressure difference p_s is very nearly equal to the fan total pressure (due to the static depression caused by the inlet velocity pressure) and thus ψ'_s may be taken as equal to ψ for practical purposes.

(b) *Pressure loss within the fan assembly.* Little accurate information seems to be available on this, but it is felt to be reasonable to consider three phases of loss.

   (i) Air enters the impeller eye usually through a reducing section from the casing inlet, and then turns through a right angle prior to entering the blade passages. The loss here may be written

$$\Delta p_i = k_i \cdot \tfrac{1}{2}\rho v_0^2 \qquad (5.24)$$

   where k_i is a loss factor probably of the order of 0·5–0·8 and v_0 is the air velocity in the impeller eye.

 (ii) Pressure loss will occur within the blade passage due to flow separation since the relative velocity of the air decreases. This loss may be written

$$\Delta p_{ii} = k_{ii} \cdot \tfrac{1}{2}\rho(w_1 - w_2)^2 \qquad (5.25)$$

   by analogy to eqn. (1.33). At the design point (that is, the point of maximum efficiency) it seems probable that k_{ii} is of the order of 0·2–0·3 for sheet metal blades, but rather less for blades of aerofoil section. At greater or lesser air flows it seems probable that increased separation occurs within the blade passage (analogous to stalling on aerofoil sections) with the effect of increasing the value of k_{ii}.

(iii) Although the casing may be designed with the intention of permitting free vortex conditions, such perfect flow is unlikely in practice. There is almost certainly an increase of area from the blade passage to the volute casing of up to about $2\frac{1}{2}$ times. There is thus a tendency of retardation of flow velocity, with resultant eddy formation. However, it is not easy to compare flow under these conditions with that at a sudden enlargement in normal pipe flow. It does seem reasonable to suppose that the pressure loss in the casing may be written

$$p_{iii} = k_{iii} \cdot \tfrac{1}{2}\rho(v_2 - v_3)^2 \qquad (5.26)$$

where v_3 is the average velocity at the fan outlet. The co-efficient k_{iii} will vary with deviation from design conditions, but at maximum efficiency is probably of the order of 0·4.

There may also be a power loss due to fluid drag on the reverse surface of the impeller backplate. If this is well spaced from the casing, the loss may be estimated since *torque* is *force × radius* which, by analogy to eqn. (1.28) may be written

$$dT = f \cdot 2\pi r \, dr \cdot \tfrac{1}{2}\rho u^2 \times r$$

where f is a friction factor of the order of 0·005 and $u = \omega r$ is the peripheral velocity of the element. On integration,

$$T = f\rho\omega^2 r^5/5$$

If the backplate is close to the casing, spaced by a distance s, it may be that viscous forces predominate and

$$dT = \mu \, dv/dy \times 2\pi r \, dr \times rn \frac{\mu \omega r}{s} 2\pi r^2 dr$$

On integration, $T = \pi\mu\omega r^4/2s$. From either of these expressions an estimate of the power loss may be made. This may be significant for large fans.

It is seen from manufacturers' data that all fans of similar type do not have the same efficiency and consequently prediction of fan performance on the above basis is likely to be less than precise. The justification for the values of k_i, k_{ii} and k_{iii} quoted is merely the degree of agreement shown in Fig. 5.7 between operating values of ψ at maximum efficiency of a number of commercial fans and calculated values based on

$$\psi_\infty = \frac{\rho u_2 v_{u2}}{\tfrac{1}{2}\rho u_2^2} = \frac{2v_{u2}}{u_2}$$

$$\psi_z = \frac{\rho u_2 (v_{u2} - \pi \sin \beta_2 u_2/z)}{\tfrac{1}{2}\rho u_2^2} = \frac{2v_{u2}}{u_2} - \frac{2\pi \sin \beta_2}{z}$$

and substituting from eqns. (5.7), (5.10) and (5.13), gives,

for $\beta_2 < 90°$ $\psi_z = 2(1 - \phi d_2 \cot \beta_2/4b_2) - 2\pi \sin \beta_2/z$,

for $\beta_2 = 90°$ $\psi_z = 2 - 2\pi \sin \beta_2/z$,

for $\beta_2 > 90°$ $\psi_z = 2(1 + \phi d_2 \cot (180° - \beta_2)/4b_2) - 2\pi \sin \beta_2/z$.

The number of blades z has been taken as 12 for $\beta_2 = 90°$ or less, and 48 for β_2 of greater than 90°, which accounts for the discontinuity in the curve.

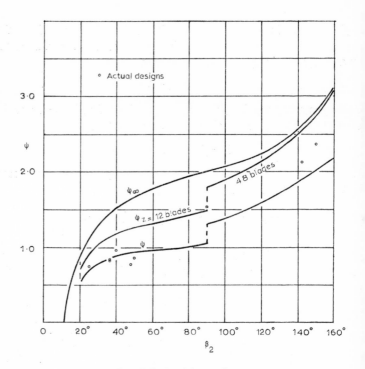

FIG. 5.7. Calculated fan performance.

Since

$$\phi = \frac{4Q}{\pi d_2^2 u_2} = \frac{4 v_{m2} \pi d_2 b_2}{\pi d_2^2 u_2} = \frac{4 b_2 v_{m2}}{d_2 u_2}$$

putting $v_{m2}/u_2 = 0.2$ gives $\phi = 0.8 b_2/d_2$.

No allowance has been made for leakage, but in the calculation of losses, $k_i = 0.8$, $k_{ii} = 0.25$ and $k_{iii} = 0.4$, and efficiency, $\eta = \psi/\psi_z$, where ψ is ψ_z less all losses. Other relationships used in the derivation of Fig. 5.7 and column J of Table 5.1 are

$$v_0 = 0.4 u_2, \quad v_3 = 0.22 u_2, \quad v_{m1} = v_{m2}$$

TABLE 5.1. Centrifugal Fan Design Details

Design	A	B	C†	D	E	F	G	H	J‡
β_2	25°	37°	37°	44°	48°	50°	142°	150°	40°
ψ	0·74	0·81	0·80	0·95	0·77	0·85	2·11	2·36	0·89
ϕ	0·19	0·15	0·14	0·19	0·27	0·14	0·40	0·67	0·18
z	12	12	12	12	12	16	60	48	12
$\phi\dfrac{d_2}{b_2}$	0·65	0·69	0·64	0·79	1·13	0·70	0·79	1·69	0·80
η	73%	77%	82%	83%	75%	70%	71%	66%	75%
v_3/u_2	0·22	0·31	0·29	0·25	0·28	0·23	0·49	1·26	0·22
v_{m2}/u_2	0·15	0·19	0·18	0·20	0·28	0·17	0·20	0·42	0·20
v_{m2}/v_{m1}	1·28	1·13	1·13	0·98	0·97	0·85	0·83	0·82	1·0
v_0/v_{m1}	2·14	2·0	2·0	1·75	1·75	1·66	2·31	1·92	2·0
d_1/d_2	0·76	0·70	0·70	0·72	0·73	0·65	0·83	0·82	0·70
b_2/d_2	0·24	0·22	0·22	0·23	0·24	0·21	0·50	0·40	0·23
$A_{\text{out}}/A_{\text{in}}$	0·78	0·48	0·48	0·78	1·0	0·85	0·78	0·48	1·0

†Aerofoil section blades, 22·7 in. in diameter.
‡Calculated values (see text).

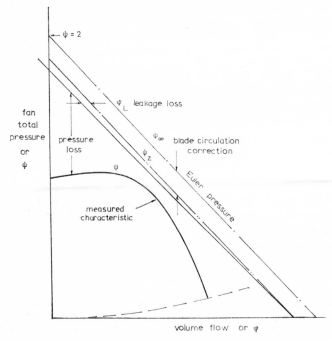

FIG. 5.8. Effect of losses in a backward curved centrifugal fan.

Figure 5.8 shows the effect of losses in a backward curved centrifugal fan on the theoretical characteristic of the impeller.

5.6. CENTRIFUGAL FAN DESIGN MODIFICATIONS

Having designed and tested a fan, the performance of geometrically similar fans of quite a large range of sizes and speeds may be predicted by means of the fan laws (eqns. (2.12) to (2.16)). These imply that, for a given operating point, the velocity triangles remain the same shape regardless of size and speed of fan. For operation at maximum efficiency, but at different ϕ/ψ ratios, it is possible, by maintaining the impeller outlet velocity triangle the same shape, to predict the performance over a moderate range for fans which are not entirely geometrically similar but which have the same blade angle. If the required fan pressure is denoted by p', and the pressure obtained from the fan tested is p, then the new peripheral velocity,

$$u'_2 = u_2 \sqrt{(p'/p)} \qquad (5.27)$$

since $p \propto u^2$ if β_2 is constant. As the angle made by the relative velocity of the air, w_2, is rather less than β_2, it is desirable to maintain a similar blade formation. This may be represented by the ratio of blade spacing to blade radial depth.

Having found the required value for u'_2, the new impeller, diameter d'_2 and/or speed n' may be fixed. To find the new impeller width,

$$v'_{m2}/v_{m2} = u'_2/u_2 \qquad (5.28)$$

and $b'_2 = Q'/\pi d'_2 v'_{m2}$, making allowance where necessary for any change in blade thickness. Since it is desirable, as far as possible, to maintain the same velocity ratios throughout:

$$v'_{m2}/v_{m2} = v'_{m1}/v_{m1} = v'_0/v_0 = u'_2/u_2 \qquad (5.29)$$

from which v'_{m1} and v'_0 may be calculated. The impeller inlet diameter d'_1 may be found since $Q' = v'_0 . \pi d'^2_1/4$, and impeller width at this diameter, b'_1, follows from a similar calculation as that for b'_2. If necessary, the number of blades should now be adjusted to main-

ain the same ratio of blade mean circumferential pitch s to blade
adial depth c:

$$\frac{s}{c} = \frac{\pi \cdot \frac{1}{2}(d_2 + d_1)}{z \cdot \frac{1}{2}(d_2 - d_1)} = \frac{\pi(d_2 + d_1)}{z(d_2 - d_1)} = \frac{\pi(d_2' + d_1')}{z(d_2' - d_1')} \tag{5.30}$$

It is unlikely to be possible to maintain the same ratio of u_1'/u_1 as
u_2'/u_2, and a new blade angle $\beta_1' = \tan^{-1}(v_{m1}'/u_1')$ must be calculated
or the impeller inlet.

The casing spiral may be made in proportion to the impeller
diameter, and the width in proportion to the impeller width.

5.7. BLADE SHAPE

The flow of air through the blade passage of a centrifugal fan
impeller is often far from ideal, and the object of design of blade

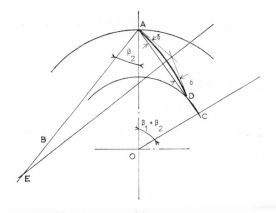

Fig. 5.9. Circular arc fan blade.

curvature should be to provide the minimum of flow separation.
This is probably best achieved in backward curved fans by having
blades of aerofoil section working at low angle of attack. However, design is still somewhat empirical. In the absence of experimental information in such a form as to permit prediction of flow
pattern to be easily made, most sheet metal blades join inlet angle β_1
to outlet angle β_2 as smoothly as possible with either a curve or a
straight line.

A circular arc is convenient to manufacture and a simple geometrical construction is given in Fig. 5.9. After laying out inner and
outer diameters of the impeller, line AB is drawn from the outer circle
at an angle β_2 to the radius OA, and line OC drawn at an angle
$\beta_1+\beta_2$ to OA to cut the inner circle at C. Line AC (extended if
necessary) will also cut the inner circle at D. Line AD is now bisected
at right angles, the bisector meeting line AB at E. With radius AE
(or DE) the circular blade profile may now be drawn. The justification for this construction is as follows.

$$\angle OAD = 90° - (\beta_2 + \delta)$$

$$\angle ODA = 90° + (\beta_1 - \delta)$$

$$\angle AOD = 180° - (\angle OAD + \angle ODA)$$

$$= 180° - (90° - \beta_2 - \delta + 90° + \beta_1 - \delta)$$

$$= \beta_2 - \beta_1 + 2\delta$$

also

$$\angle ODC = 180° - \angle ODA = 180° - (90° + \beta_1 - \delta)$$

$$= 90° - \beta_1 + \delta$$

and

$$\angle DOC = 180° - 2\angle ODC = 180° - (180° - 2\beta_1 + 2\delta)$$

$$= 2\beta_1 - 2\delta$$

$$\angle AOC = \angle AOD + \angle DOC = \beta_2 - \beta_1 + 2\delta + 2\beta_1 - 2\delta$$

$$= \beta_2 + \beta_1$$

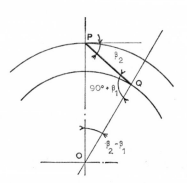

FIG. 5.10. Straight fan blade.

For straight backward blades, the construction is rather simpler, but it is not always possible to obtain all design requirements. The outside circle at diameter d_2 is set out, and line PQ is drawn at an angle of $90° - \beta_2$ to the radial line OP to meet line OQ at $\beta_2 - \beta_1$ to OP. Point Q is then the correct point on the inner circle. With this shape of blade it is unlikely that radius OQ will be the same as the design radius and a compromise must be made between β_1 and d_1. However, such blade shapes appear to perform satisfactorily in practice.

5.8. AXIAL FLOW FAN VELOCITY TRIANGLES

Figure 5.11 shows an axial flow fan blade section at some particular radius, with its associated velocity triangles. The air enters the impeller axially with a velocity $v_1 = v_{m1}$, and leaves with velocity v_2. The shape of the triangles is almost identical with those of a backward bladed centrifugal fan, but it should be noted that $u_1 = u_2$, and

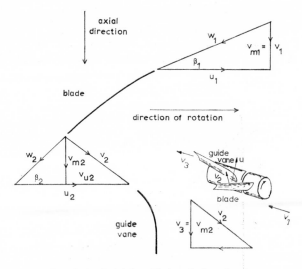

FIG. 5.11. Axial flow blade velocity triangles.

$v_{m1} = v_{m2}$. The total pressure developed is given by the same equation as for a centrifugal fan (eqn. (5.4)), namely, $\rho u_2 v_{u2}$, v_{u2} being the rotational component of v_2. It should be noted that the expanded form of Euler's equation (eqn. (5.3)) no longer includes a forced vortex component since $u_1 = u_2$ at each radius.

The theoretical characteristics may be derived since:

$$v_u = u - v_m \cot \beta_2$$

$$p = \rho u v_u = \rho u^2 - \rho u v_m \cot \beta_2$$

$$= \rho u^2 - \rho u \cdot \frac{4Q}{\pi d_2^2 (1-v^2)} \cdot \cot \beta_2 \qquad (5.31)$$

where $v = d_1/d_2$, or, in terms of non-dimensional coefficients,

$$\psi = 2(1 - \phi' \cot \beta_2) \qquad (5.32)$$

where

$$\phi' = v_m/u = 4Q/\pi d_2^2 (1-v^2)u = \phi/(1-v^2) \qquad (5.33)$$

The flow coefficient ϕ' is useful for the design of a blade element. Power coefficient,

$$\lambda = \phi\psi = 2\phi\left[1 - \frac{\phi}{(1-v^2)} \cdot \cot \beta_2 \right] \qquad (5.34)$$

The characteristics are shown in Fig. 5.12, and are seen to be almost identical with those for a backward bladed centrifugal fan.

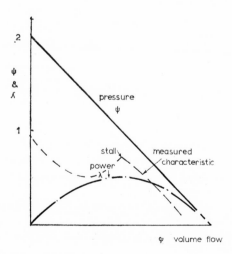

Fig. 5.12. Theoretical characteristics of an axial flow impeller.

It is common to design a blade to give the same axial velocity and pressure development at each radius, in which case $p = \rho \omega r v_u =$ constant, or $r v_u =$ constant. This will be seen to be the condition for a free vortex and permits radial equilibrium of forces on the fluid. It is necessary to have increased blade angles at the hub section to achieve the higher values of v_u at the smaller radius. Departures from free vortex designs have been made.

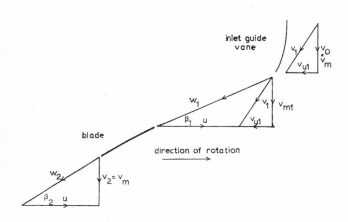

FIG. 5.13. Axial flow blade with upstream guide vane.

Since the air leaving the impeller has a rotational component of velocity, v_u, there is a loss of total pressure of

$$\tfrac{1}{2}\rho(v_2^2 - v_m^2) = \tfrac{1}{2}\rho v_u^2 \qquad (5.35)$$

if the rotational energy is allowed to be dissipated along the duct system. Downstream guide vanes may be fitted as shown in Fig. 5.11 to reduce the velocity to v_m and thereby regain static pressure equal to $\tfrac{1}{2}\rho v_u^2$. Even so, many commercial designs are produced without guide vanes to reduce costs, the resulting loss in efficiency being relatively unimportant at low fan power.

It is possible to avoid rotational energy loss by having a guide vane upstream of the impeller which pre-rotates the entering air in a direction opposite to that of the impeller rotation. The impeller is designed to do sufficient work on the air to remove this rotation (Fig. 5.13). Then, from eqn. (5.2),

$$p = \rho u_2 v_{u2} - \rho u_1 v_{u1} = 0 - \rho u_1(-v_{u1})$$

$$= \rho u v_{u1} \qquad\qquad (5.36)$$

and, at the design point,

$$v_{u1} = v_m \cot \beta_1 - u$$

$$\rho u v_{u1} = \rho u v_m \cot \beta_1 - \rho u^2$$

or

$$\psi = 2(\phi' \cot \beta_1 - 1) \qquad\qquad (5.37)$$

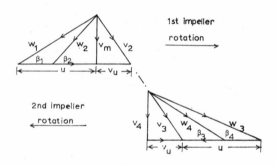

FIG. 5.14. Contra-rotating fan velocity triangles.

A third type of unit, the contra-rotating fan, makes use of air leaving an impeller with rotation to enter a second impeller rotating in the opposite direction. This second impeller acts in a similar manner to that of an upstream guide vane fan, as can be seen from the velocity triangles. In Fig. 5.14 the inlet and outlet velocity triangles for each impeller have been combined into a single diagram made possible since v_m and u are the same in each case. Each impeller develops the same pressure if u and v_u for each are the same, and the air is discharged axially, that is

$$p = 2\rho u v_u \qquad\qquad (5.38)$$

A similar arrangement, with both impellers running in the same direction, is possible by using guide vanes between the impellers

lowever, a large angular deflection of the air is necessary and care-
ul design of the guide vanes is essential to avoid flow separation.

5.9. USE OF AEROFOIL SECTION BLADES

As with centrifugal fans, the air passing through an impeller
onstructed with sheet metal blades will not follow the blade profile

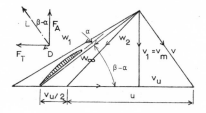

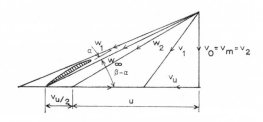

FIG. 5.15. Use of aerofoil section axial flow blades.

very accurately unless the number of blades is infinite. Since aerofoil
data are available, it is possible to predict the performance of an
axial flow fan more accurately if blades of aerofoil profile are used.
The velocity triangles for such blades are shown in Fig. 5.15 and are
seen to differ from those previously considered only by the addition
of a mean relative velocity vector, $w_\infty = \frac{1}{2}(w_1 + w_2)$ to which the
blade section is inclined at its angle of attack, α. The mean blade
angle is β, with an *effective* blade angle (blade air angle) between
vectors of w and u of $\beta - \alpha$.

The static pressure difference across the impeller may be found since

$$p = \rho u v_u = p_{t1} - p_{t2} = p_2 + \tfrac{1}{2}\rho v_2^2 - (p_1 + \tfrac{1}{2}\rho v_1^2)$$
$$= p_2 - p_1 + \tfrac{1}{2}\rho(v_2^2 - v_1^2)$$

Static pressure difference,

$$p_2 - p_1 = \rho u v_u - \tfrac{1}{2}\rho(v_2^2 - v_1^2)$$
$$= \rho u v_u \pm \tfrac{1}{2}\rho v_u^2$$
$$= \rho v_u(u \pm \tfrac{1}{2}v_u) \qquad (5.39)$$

where the negative sign refers to the downstream guide vane impeller and the positive sign to the upstream guide vane impeller.

This pressure difference over the impeller swept area may be equated to the axial thrust due to the aerodynamic lift forces L on the blades (eqn. (1.62)),

$$F_A = L\cos(\beta - \alpha) = (p_2 - p_1).2\pi r dr$$

for an element of blade. If there are z blades, each of chord c,

$$zc.dr.C_L.\tfrac{1}{2}\rho w_\infty^2 \cos(\beta - \alpha) = \rho v_u(u \pm \tfrac{1}{2}v_u)2\pi r.dr$$

and writing blade spacing, $s = 2\pi r/z$

and substituting

$$u \pm \tfrac{1}{2}v_u = w_\infty \cos(\beta - \alpha)$$
$$\tfrac{1}{2}\frac{c}{s}.C_L w_\infty = v_u \qquad (5.40)$$

or

$$\frac{v_u}{w_\infty} = \tfrac{1}{2}C_L\frac{c}{s}$$

The above simplified blade element theory, whilst probably adequate for exploratory design, ignores the effect of drag. To consider more fully the forces on the aerofoils it is necessary to equate the thrust force F_A, which is due to static pressure rise less any pressure loss, to the axial force due to the lift and drag. Also the tangential force F_T due to lift and drag must be equated to the rate of change of tangential momentum, thus

$$F_A = (p_s - \Delta p)2\pi r \cdot dr = L\cos(\beta - \alpha) - D\sin(\beta - \alpha) \tag{5.41}$$

$$F_T = \rho v_m v_u 2\pi r \cdot dr = L\sin(\beta - \alpha) + D\cos(\beta - \alpha) \tag{5.42}$$

Expanding eqn. (5.42),

$$\rho v_m v_u 2\pi r \cdot dr = zc \cdot dr \cdot \tfrac{1}{2}\rho w_\infty^2 [C_L\sin(\beta - \alpha) + C_D\cos(\beta - \alpha)]$$

and since $w\sin(\beta - \alpha) = v_m$ and $s = 2\pi r/z$,

$$v_u\sin(\beta - \alpha) = \tfrac{1}{2}\frac{c}{s}w_\infty [C_L\sin(\beta - \alpha) + C_D\cos(\beta - \alpha)]$$

$$\frac{v_u}{w_\infty} = \frac{C_L' c}{2s} \tag{5.43}$$

where

$$C_L' = C_L\left[1 + \frac{C_D}{C_L}\cot(\beta - \alpha)\right] \tag{5.44}$$

Expanding eqn. (5.41),

$$[\rho v_u(u \pm \tfrac{1}{2}v_u) - \Delta p] \cdot 2\pi r \cdot dr$$

$$= zc \cdot dr \tfrac{1}{2}\rho w_\infty^2 [C_L\cos(\beta - \alpha) - C_D\sin(\beta - \alpha)]$$

and since $u + \tfrac{1}{2}v_u = w \cdot \cos(\beta - \alpha)$,

$$\Delta p = \rho v_u w_\infty \cos(\beta - \alpha) - \tfrac{1}{2} \cdot \frac{c}{s} \cdot \rho w_\infty^2 [C_L\cos(\beta - \alpha) - C_D\sin(\beta - \alpha)]$$

Now,

$$v_u w_\infty = v_u w_\infty^2/w_\infty = w_\infty^2 \tfrac{1}{2}\frac{c}{s}[C_L + C_D\cot(\beta - \alpha)]$$

thus

$$\Delta p = \tfrac{1}{2}\rho w_\infty^2 \cdot \frac{c}{s}\left[C_L\cos(\beta - \alpha) + \frac{C_D\cos^2(\beta - \alpha)}{\sin(\beta - \alpha)} - C_L\cos(\beta - \alpha)\right.$$

$$\left. + C_D\sin(\beta - \alpha)\right]$$

$$= \tfrac{1}{2}\rho w_\infty^2 \cdot \frac{c}{s} \cdot C_D\left[\frac{\cos^2(\beta - \alpha) + \sin^2(\beta - \alpha)}{\sin(\beta - \alpha)}\right]$$

$$\Delta p = \tfrac{1}{2}\rho w_\infty^2 \cdot \frac{c}{s} \cdot C_D\operatorname{cosec}(\beta - \alpha) \tag{5.45}$$

For the Göttingen 436 aerofoil section, the practical details of which are given in Fig. 1.13 (for a model having a 3 in. chord) C_D/C_L has a value of less than $\frac{1}{60}$ for a value of C_L of unity. This occurs at an angle of attack of a little less than 6°, for which the value of C_D is about 0·016. The blade air angle, $\beta - \alpha$, is usually between about 10° and 40° and may be found from the velocity triangles as

$$\tan^{-1}v_m/(u+\tfrac{1}{2}v_u).$$

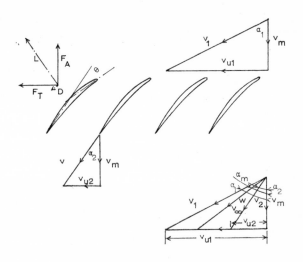

FIG. 5.16. Cascade of aerofoils.

Values of C_L in the above equations will normally be based on wind tunnel tests on single aerofoils of infinite aspect ratio, since an axial flow fan blade is bounded by hub and duct. The effect of increased tip clearance is to reduce the performance of the fan, and to increase losses at the tips of the blade. The effect of a finite number of blades is to modify the flow over each aerofoil due to interference by adjacent blades. These effects appear not to be very serious for values of c/s not exceeding unity, but for closer blade spacings it is advisable to make use of cascade data from wind tunnel tests on cascades or grids of aerofoils.[20]

In the cascade shown in Fig. 5.16, the mean absolute velocity vector is taken as $v_\infty = \tfrac{1}{2}(v_1+v_2)$, and the flow angles α_1 and α_2, and

ow velocities are measured values from the tests. Forces may be
quated as for eqns. (5.41) and (5.42).

$$F_A = L \sin \alpha_m - D \cos \alpha_m \qquad (5.46)$$

$$F_T = L \cos \alpha_m + D \sin \alpha_m \qquad (5.47)$$

Now, $F_A = s[(p_2 - p_1) - \Delta p]$, where Δp is the pressure loss

$$= s[\tfrac{1}{2}\rho(v_1^2 - v_2^2) - \Delta p] \qquad (5.48)$$

and

$$F_T = s\rho v_m(v_{u1} - v_{u2}) = s\rho v_m^2(\tan \alpha_1 - \tan \alpha_2) \qquad (5.49)$$

From eqns. (5.47) and (5.49)

$$s\rho v_m^2(\tan \alpha_1 - \tan \alpha_2) = c \cdot \tfrac{1}{2}\rho v_\infty^2(C_L \cos \alpha_m + C_D \sin \alpha_m),$$

$$s\rho v_\infty^2 \cos^2 \alpha_m(\tan \alpha_1 - \tan \alpha_2) = c \cdot \tfrac{1}{2}\rho v_\infty^2(C_L \cos \alpha_m + C_D \sin \alpha_m),$$

$$2 \cdot \frac{s}{c} \cos \alpha_m(\tan \alpha_1 - \tan \alpha_2) = C_L + C_D \tan \alpha_m$$

or

$$C_L = 2 \cdot \frac{s}{c}(\tan \alpha_1 - \tan \alpha_2) \cos \alpha_m - C_D \tan \alpha_m \qquad (5.50)$$

Considering the axial forces,

$$\tfrac{1}{2}\rho(v_1^2 - v_2^2) = \tfrac{1}{2}\rho(v_m^2 + v_{u1}^2 - v_m^2 - v_{u2}^2)$$

$$= \tfrac{1}{2}\rho(v_{u1}^2 - v_{u2}^2)$$

$$= \tfrac{1}{2}\rho v_m^2(\tan^2 \alpha_1 - \tan^2 \alpha_2)$$

$$= \tfrac{1}{2}\rho v_m^2(\tan \alpha_1 + \tan \alpha_2)(\tan \alpha_1 - \tan \alpha_2)$$

But,

$$\tan \alpha_m = \frac{v_\infty \sin \alpha_m}{v_m} = \tfrac{1}{2}(v_{u1} + v_{u2})/v_m = \tfrac{1}{2}(\tan \alpha_1 + \tan \alpha_2) \qquad (5.51)$$

Thus $\tfrac{1}{2}\rho(v_1^2 - v_2^2) = \rho v_m v_\infty \sin \alpha_m(\tan \alpha_1 - \tan \alpha_2)$
and eqn. (5.48) becomes, on substitution,

$$F_A = s[\rho v_\infty^2 \cos \alpha_m \sin \alpha_m(\tan \alpha_1 - \tan \alpha_2) - \Delta p]$$

$$= c \cdot \tfrac{1}{2}\rho v_\infty^2(C_L \sin \alpha_m - C_D \cos \alpha_m) \text{ from eqn. (5.46),}$$

$$= c \cdot \tfrac{1}{2}\rho v_\infty^2\left[\frac{2s}{c}(\tan \alpha_1 - \tan \alpha_2) \cos \alpha_m \sin \alpha_m - C_D \tan \alpha_m \sin \alpha_m\right.$$

$$\left. - C_D \cos \alpha_m\right]$$

on substitution of the value of C_L from eqn. (5.50), and

$$\Delta p = \frac{c}{s}. C_D(\tan \alpha_m \sin \alpha_m + \cos \alpha_m). \tfrac{1}{2}\rho v_\infty^2$$

$$= \frac{c}{s}. \frac{C_D}{\cos \alpha_m} (\sin^2 \alpha_m + \cos^2 \alpha_m). \tfrac{1}{2}\rho v_\infty^2$$

$$= \frac{c}{s}. \frac{C_D}{\cos \alpha_m}. \tfrac{1}{2}\rho v_\infty^2 \qquad (5.52)$$

or

$$C_D = \frac{s}{c}. \frac{\cos \alpha_m}{\tfrac{1}{2}\rho v_\infty^2}. \Delta p = \frac{s}{c} \frac{\cos^3 \alpha_m}{\cos^2 \alpha_1}. \frac{\Delta p}{\tfrac{1}{2}\rho v_1^2} \qquad (5.53)$$

since $v_m = v_1 \cos \alpha_1 = v_\infty \cos \alpha_m$.

Thus the values of C_L and C_D may be calculated. With complete cascade data, however, it is probably simpler to use the deflection directly since, for a fan blade element,

$$v_u = v_m(\tan \alpha_1 - \tan \alpha_2)$$

and total pressure,

$$p = \rho u v_u = \rho u v_m(\tan \alpha_1 - \tan \alpha_2) \qquad (5.54)$$

5.10. LOSSES IN AXIAL FLOW FANS

Pressure loss in axial flow fans occurs in two main sections; the blades and the guide vanes. The loss in the blades due to profile drag is given in eqn. (5.45). Other losses are due to " friction " at the duct wall and the hub, and secondary losses due to trailing vortices. The use has been suggested[20] of an overall drag coefficient,

$$C'_D = C_D + C_{Da} + C_{Ds} + C_{Dt} \qquad (5.55)$$

where

C_D = normal aerofoil profile drag coefficient,

C_{Da} = annulus drag coefficient

$\quad = 0.02\, s/h$, where h is the blade length, $\qquad (5.56)$

C_{Ds} = secondary drag loss coefficient

$\quad = 0.018\, C_L^2,$ $\qquad (5.57)$

C_{Dt} = tip drag loss $= 0.029 . \dfrac{\delta}{h}. C_L^{\frac{3}{2}}$ $\qquad (5.58)$

where δ is the tip clearance.

Similar losses occur in the guide vane section. Where these are of aerofoil section, losses will be similar to those of the blade and may be based on a value of w_∞ which is the mean of v_2 (or v_0) and v_m. For sheet metal guide vanes, the loss should be of a similar order to that in a straight expander of the same effective angle.

5.11. THE DESIGN OF AXIAL FLOW FANS

Since the magnitude of tangential velocity u is least at the hub section, and consequently v_u is a maximum, it is desirable to start

TABLE 5.2. Axial Flow Fan Details

Design source	ν	ϕ	ψ	η	
A1	0·33	0·198	0·044	78%	⎫
A2	0·33	0·180	0·086	79%	⎬ upstream guide vanes
A3	0·47	0·230	0·16	84%	
A4	0·7	0·365	0·50	77%	
B	0·5	0·25	0·30	85%	⎭
C1	0·45	0·277	0·38	86%	⎫ downstream guide vanes
C2	0·5	0·268	0·52	85%	⎬
D1	0·5	0·244	0·155	78%	⎫ no guide vanes
D2	0·25	0·17	0·11	84%	⎬
E	0·55	0·145	0·25	86%	⎫ downstream guide vanes
F	0·55	0·225	0·24	79%	⎬
G	0·50	0·203	0·49	75%	contra-rotating

the design of an axial flow fan at this radius. This means that both hub and tip tangential velocities must first be settled, and on this point, experience of previous design will help to reduce trial and error attempts. As a guide, the values in Table 5.2 for actual fans may be used as a starting point. Speeds of rotation will usually be predetermined by those available on a.c. squirrel cage induction motors, since it is convenient to have impellers driven directly by the motor shaft.

The value of u at the hub may now be calculated and the required value of v_u ($= p/\rho u$) found. Since tip and hub diameters are now known, v_m may be calculated [$= 4Q/\pi(d_2^2 - d_1^2)$]. It is next possible to find v_u/w_∞, since $w = \sqrt{[v_m^2 + (u + \frac{1}{2}v_u)^2]}$, the value of which should not exceed 0·5–0·6. This is because C_L' will generally have a maximum usable value of 1·0–1·2 prior to stall, and c/s should not exceed unity in a preliminary design. If the value of v_u/w_∞ exceeds this value the diameter of the hub must be increased until practical values can be achieved. The blade air angle, $\beta - \alpha = \tan^{-1}[v_m/(u \pm \frac{1}{2}v_u)]$. The actual blade angle β is found by adding to $\beta - \alpha$ the value of α appropriate to the value of C'_L selected.

It should now be a straightforward matter to design the sections at other radii between hub and tip. Since it is undesirable to have a

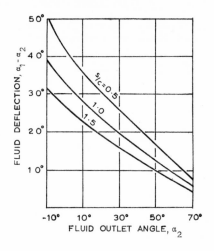

FIG. 5.17. Cascade performance for fluid deflections which are 80% of stall deflections[20] by aerofoils having a camber angle of about 25–30°.

considerable rotating mass in the impeller, the solidity will generally reduce from hub to tip. Blade chord will tend to remain constant or reduce somewhat towards the tip, consistent with maintenance of accuracy of blade profile (which will often be selected with convenience of manufacture in mind).

Having designed the impeller, the guide vanes may now be designed to give the correct flow deflections at the operating point, preferably from cascade data (Fig. 5.17). In order to avoid excessive loss between the blade and guide vane hub section and the fan discharge, a fairing permitting a gradual increase in area and tapering at an angular rate of not more than 15° total angle for a length sufficient to give reasonable static pressure regain is advisable. This is particularly important for fans with large hubs where an increase of area of airway of the order of two to one should be possible to obtain in a relatively short length. Where fans are to be used with free entry, it is also necessary to give good entry conditions by the addition of an elliptical inlet nozzle, or at least a 60° included angle cone of a length of about one-quarter of the impeller diameter.

5.12. EXAMPLES

In these examples, air is taken as having a density of 0·075 lb/ft³.

1. A backward curved fan impeller has a blade angle β_2 of 40°. The radial air velocity is 30 ft/s and the impeller peripheral velocity is 100 ft/s. Find the theoretical head developed by the impeller.

From Fig. 5.2 it is seen that $u_2 - v_{u2} = v_m \cot \beta_2$.

$v_{u2} = u_2 - v_m \cot \beta_2 = 100 - 30 \cot 40° = 100 - 30 \times 1·192$
$\quad = 100 - 35·8 = 64·2$ ft/s.

For normal inlet conditions, the theoretical head developed by the impeller,

$$H = \frac{u_2 v_{u2}}{g} \quad \text{(see eqn. (5.4))}$$

$$= \frac{100 \times 64·2}{32·2} = 199 \text{ ft.}$$

Since this represents the height to which air must be raised to give the same energy per unit weight as the work done by the fan impeller, the equivalent head of water is easily found if desired since

$w_w H_w = w_a H_a$, and taking w_w as 62·4 lb/ft³,

$$H_w = \frac{w_a H_a}{w_w} = \frac{0·075 \times 199}{62·4} = 0·239 \text{ ft, or } 2·87 \text{ in.}$$

2. Calculate, making corrections for interblade circulation, the impeller outer and inner diameters and blade widths, and inlet blade angle of a backward bladed centrifugal fan having 12 blades of angle β_2 of 50° for a duty of 4000 ft³/min at a fan static pressure of 3·0 in. of water. The fan total efficiency may be taken as 75%, all losses being pressure losses, and the casing as having an outlet area of 2 ft². Other data:

radial air velocity in the impeller, $v_m = 0·2\, u_2$,
air velocity at impeller inlet (eye), $v_0 = 0·4\, u_2$,
speed of rotation of impeller, $n = 900$ rev/min.

Correction for interblade circulation $= \dfrac{\pi \sin \beta_2}{z} \times u_2$ (eqn. (5.15))

$$= \frac{\pi}{12} \times \sin 50° \times u_2 = \frac{\pi}{12} \times 0·766 \times u_2 = 0·2\, u_2.$$

$v_{u2} = v_{u2} - 0·2\, u_2 = u_2 - v_m \cot 50° - 0·2\, u_2$
$\quad = 0·8\, u_2 - 0·2\, u_2 \cot 50° = 0·8\, u_2 - 0·2\, u_2 \times 0·839$
$\quad = 0·632\, u_2.$

Impeller total pressure $= \rho u_2 v_{u2} = 0 \cdot 075 \times u_2 \times 0 \cdot 632\, u_2$
$$= 0 \cdot 0473\, u_2^2.$$

Fan outlet velocity $= 4000/2 = 2000$ ft/min.

Fan velocity pressure $= (2000/4000)^2 = 0 \cdot 25$ in. water.

Fan total pressure $= 3 \cdot 0 + 0 \cdot 25 = 3 \cdot 25$ in. water.

At 75% total efficiency, the required impeller total pressure will be

$$3 \cdot 25/0 \cdot 75 = 4 \cdot 33 \text{ in. water.}$$

In ft lb s units this is $4 \cdot 33 \times 5 \cdot 2 \times 32 \cdot 2 = 725$ pdl/ft². Equating the two values for impeller total pressure, $725 = 0 \cdot 0473\, u_2^2$, from which

$$u_2 = \sqrt{(725/0 \cdot 0473)} = \sqrt{15,320} = 123 \cdot 6 \text{ ft/s.}$$

Impeller diameter $= \dfrac{123 \cdot 6 \times 60}{\pi \times 900} = 2 \cdot 63$ ft.

Velocity through impeller eye $= 0 \cdot 4 \times 123 \cdot 6 = 49 \cdot 4$ ft/s.

Area of impeller eye $= \dfrac{4000}{60 \times 49 \cdot 4} = 1 \cdot 35$ ft².

Diameter of impeller eye $= \sqrt{(4 \times 1 \cdot 35/\pi)} = 1 \cdot 32$ ft.

Now, $Q = \pi d_2 b_2 v_m = \pi d_1 b_1 v_m$, where $v_m = 0 \cdot 2 \times 123 \cdot 6 = 24 \cdot 7$ ft/s,

$$b_2 = \frac{Q}{\pi d_2 v_m} = \frac{4000}{\pi \times 2 \cdot 63 \times 24 \cdot 7 \times 60} = 0 \cdot 327 \text{ ft,}$$

$$b_1 = \frac{Q}{\pi d_1 v_m} = \frac{4000}{\pi \times 1 \cdot 32 \times 24 \cdot 7 \times 60} = 0 \cdot 65 \text{ ft.}$$

Inlet blade angle, $\beta_1 = \tan^{-1}(v_m/u_1) = \tan^{-1}\left(\dfrac{24 \cdot 7 \times 3 \cdot 63}{123 \cdot 6 \times 1 \cdot 32}\right)$

$$= \tan^{-1}(0 \cdot 40) = 21 \cdot 8°.$$

3. A centrifugal fan has the following details:

Impeller; outside diameter, 27 in., inside diameter, 20·5 in.:

blade widths, 6·5 and 11 in. respectively:

blade angles, $\beta_2 = 25°$, $\beta_1 = 31°$:

number of blades, 12:

casing outlet, 26 in. high by 17 in. wide.

When the impeller rotates at 960 rev/min, the fan has a duty of 4930 ft³/min at 1·80 in. of water fan static pressure. A fan is required for a duty of 4000 ft³/min at 1·80 in. of water fan total pressure. It is hoped to use the same casing with the minimum of modification, and it is to run at the same speed as the original fan. Design an impeller to meet these requirements.

Original fan outlet velocity $= \dfrac{4930 \times 144}{26 \times 17} = 1610$ ft/min.

Fan velocity pressure $= (1610/4000)^2 = 0 \cdot 16$ in. water.

Original fan total pressure $= 1 \cdot 80 + 0 \cdot 16 = 1 \cdot 96$ in. water.

Maintaining the same blade angle, β_2, and, as far as possible, the same blade spacing, the laws of similarity may be applied to the velocity vectors in the fan velocity triangles.

Fan total pressure $\propto uv_u \propto u^2 \propto (\pi nd)^2 \propto d^2$.

Thus new diameter $= 27\sqrt{(1\cdot8/1\cdot96)} = 25\cdot9$ in., and the ratio of the velocities will be $\sqrt{(1\cdot8/1\cdot96)} = 0\cdot96$.

Velocity through impeller eye, $v_0 = 4Q/(\pi d_1^2)$

$v_0'/v_0 = 0\cdot96 = (Q'/d_1'^2)\times(d_1^2/Q) = (4000/4930)\times(20\cdot5/d_1)^2$

$\qquad d_1' = 20\cdot5\sqrt{[4000/(4930\times0\cdot96)]} = 18\cdot85$ in.

Impeller radial velocity, $v_{m2} = Q/(\pi d_2 b_2)$.

$$v_{m2}'/v_{m2} = 0\cdot96 = Q'/(d_2'b_2')\times(d_2b_2/Q) = \frac{4000\times27\times6\cdot5}{4930\times25\cdot9\times b_2'}$$

from which $b_2' = 5\cdot72$ in. Similarly, impeller inlet width may be calculated,

$$b_1' = \frac{Q_1'd_1b_1v_{m1}}{Q_1d_1'b_1'v_{m1}'} = \frac{4000\times20\cdot5\times11}{4930\times18\cdot85\times0\cdot96} = 10\cdot1 \text{ in.}$$

$$v_{m1}'/u_1' = Q/(\pi d_1'b_1'\times\pi d_1'n) = \frac{4000\times12\times12\times12}{\pi^2\times18\cdot85^2\times10\cdot1\times960} = 0\cdot202,$$

$\beta_1 = \tan^{-1}(v_{m1}'/u_1') = \tan^{-1}(0\cdot202) = 11\cdot5°$.

Checking blade spacing, $\pi(d_2+d_1)/z(d_2-d_1)$ should remain constant, thus

$$z' = \frac{(d_2'+d_1')\,(d_2-d_1)}{(d_2'-d_1')\,(d_2+d_1)}\times z$$

$$= \frac{(25\cdot9+18\cdot9)\,(27-20\cdot5)}{(25\cdot9-18\cdot9)\,(27+20\cdot5)}\times12 = \frac{44\cdot8\times6\cdot5}{7\cdot0\times47\cdot5}\times12 = 10\cdot5.$$

It is likely that 10 blades will be satisfactory.

4. A downstream guide vane axial flow fan element is to be designed for $\psi = 0\cdot2$ and $\phi^* = 0\cdot5$. What is the blade angle β if the angle of attack of the section is $5°$?

$\psi = 0\cdot2 = \rho uv_u/\tfrac{1}{2}\rho u^2 = 2v_u/u; \quad v_u = 0\cdot1u.$

$\phi^* = 0\cdot5 = v_m/u; \quad v_m = 0\cdot5u.$

$\beta-\alpha = \tan^{-1}(v_m/(u-\tfrac{1}{2}v_u)) = \tan^{-1}[0\cdot5u/(u-0\cdot05u)]$

$\qquad = \tan^{-1}(0\cdot5/0\cdot95) = \tan^{-1}(0\cdot527) = 27\cdot8°.$

$\beta = (\beta-\alpha)+\alpha = 27\cdot8°+5° = 32\cdot8°.$

5. Calculate hub and tip blade angles and guide vane angles for an upstream guide vane axial flow fan which has hub and tip diameters of 1 ft and 2 ft respectively rotating at 960 rev/min, and which has a duty of 3600 ft³/min at a fan static pressure of 0·52 in. of water, and a fan total efficiency of 85%. The following aerofoil data is available:

angle of attack of section: 0°   2°   4°   6°   8°   10°   12°
coefficient of lift: 0·2   0·4   0·6   0·8   1·0   1·2   1·4 (stall).
Blade solidity at the hub is 1·0, and at the tip is 0·4.

$$\text{Fan outlet velocity pressure} = \left(\frac{3600 \times 4}{\pi \times 4 \times 4000}\right)^2 = 0·08 \text{ in. water.}$$

Fan total pressure = $0·52 + 0·08 = 0·60$ in. water.
Required impeller total pressure = $0·60 \times 5·2 \times 32·2/0·85 = 118·5$ pdl/ft^2.
This must be equated with $\rho u v_u$.

Hub section: $u = \pi \times 1 \times 960/60 = 50·3$ ft/s.
$$v_u = 118·5/(0·075 \times 50·3) = 31·4 \text{ ft/s.}$$
$v_m = 3600 \times 4 \div [60 \times \pi \times (2^2 - 1^2)] = 25·5$ ft/s.
$\beta - \alpha = \tan^{-1}(v_m/(u + \frac{1}{2}v_u)) = \tan^{-1}(25·5/(50·3 + 15·7))$
        $= \tan^{-1}(25·5/66·0) = \tan^{-1}(0·387) = 21·2°.$
$w_\infty = \sqrt{(25·5^2 + 66·0^2)} = \sqrt{5006} = 70·8$ ft/s.
$v_u/w_\infty = 31·4/70·8 = 0·443 = \frac{1}{2}C_L c/s$ (from eqn. (5.43)).
For $c/s = 1·0$, $C_L = 0·886$ and $\alpha = 6·86°$ from given data above.
Blade angle at the hub, $\beta = 21·2° + 6·86° = 28·06°.$

Tip section: $u = \pi \times 2 \times 960/60 = 100·5$ ft/s.
$v_u = 118·5/(0·075 \times 100·5) = 15·7$ ft/s: v_m remains as before.
$\beta - \alpha = \tan^{-1}[25·5/(100·5 + 7·9)] = \tan^{-1}(25·5/108·4) = 13·2°.$
$w_\infty = \sqrt{(25·5^2 + 108·4^2)} = \sqrt{12{,}390} = 111·3$ ft/s.
$v_u/w_\infty = 15·7/111·3 = 0·141 = \frac{1}{2}C_L c/s.$
For $c/s = 0·4$, $C_L = 0·282/0·4 = 0·705$, for which $\alpha = 5·05°.$
Blade angle at the tip, $\beta = 13·2° + 5·05° = 18·25°.$
Angle of guide vane to fan axis adjacent to impeller $= \tan^{-1}(v_u/v_m).$
At the hub this is $\tan^{-1}(31·4/25·5) = 50·9°$, and at the tip the corresponding
angle is $\tan^{-1}(15·7/25·5) = 31·6°.$

6. Tests on a cascade of aerofoils, for which $c/s = 1·5$, yielded the following
data: $\alpha_1 = 50°$, $\alpha_2 = 20°$, inlet air velocity = 80 ft/s, and the pressure loss was
0·061 in. of water. Calculate values of C_L and C_D.

Tan $\alpha_m = \frac{1}{2}(v_{u1} + v_{u2})/v_m = \frac{1}{2}(\tan \alpha_1 + \tan \alpha_2)$ (see Fig. 5.16)
       $= \frac{1}{2}(\tan 50° + \tan 20°) = \frac{1}{2}(1·192 + 0·364) = 0·778.$
$\alpha_m = 37°53'$ and $\cos \alpha_m = 0·788$, giving $\cos^3 \alpha_m = 0·489.$

Inlet velocity pressure, $\frac{1}{2}\rho v_1^2$, in inches of water $= \left(\frac{80 \times 60}{4000}\right)^2$

that is, 1·44 in. From eqn. (5.53),

$$C_D = \frac{s \cos^3 \alpha_m \Delta p}{c \cos^2 \alpha_1 \frac{1}{2}\rho v_1^2} = \frac{0·489 \times 0·061}{1·5 \times 0·643^2 \times 1·44} = 0·034.$$

From eqn. (5.50),

$$C_L = 2 \frac{s}{c}(\tan \alpha_1 - \tan \alpha_2) \cos \alpha_m - C_D \tan \alpha_m$$

$$= \frac{2}{1·5}(1·192 - 0·364)0·788 - 0·034 \times 0·778$$

$$= 0·868 - 0·026 = 0·842.$$

CHAPTER 6

MECHANICAL DESIGN OF FANS

6.1. STRESS AND STRAIN OF AN ELASTIC MATERIAL

Mechanical design of fans is largely a matter of using, for constructional purposes, available materials within their limits of strength, rigidity and durability. Much of the study of stresses and strains in materials is based on principles of mechanics, backed by experimental results. In studying a section of material it is customary to consider the forces to one side of the section only, since these are usually balanced by equal and opposite forces on the other side of the section. A material in which there is action and reaction across a surface is said to be in a state of stress. Where the distribution is uniform, the force transmitted per unit area is the intensity of stress (usually referred to simply as stress). Two particular forms of stress are

(i) direct stress f where the surface stressed is normal to the line of action of the force causing it. A direct stress tending to elongate a material is known as a tensile stress, and regarded algebraically as positive, whilst one tending to shorten the piece of material is known as a compressive stress and is regarded algebraically as negative.

(ii) shear stress q where the stress is tangential to the surface considered.

A state of stress causes distortion to the original shape of a piece of material. Tensile stress causes an elongation, and the ratio extension δl to original length l is known as direct strain e. Concurrent with the direct strain there is indirect strain in directions at right angles, but of the opposite sense, in this case shortening the dimensions. The ratio indirect strain/direct strain is known as Poisson's ratio, σ, or perhaps more correctly in an algebraic sense, $-\sigma$. The indirect strain is thus $-\sigma\delta l/l$ when a tensile stress is applied. When a compressive stress is applied, there is a direct strain of $-\delta l/l$ and an indirect strain of $\sigma \delta l/l$. Shear stress causes an angular distortion ϕ, which is known as shear strain.

An elastic material is one in which the relationship between stress and strain is linear. In practical cases, limits exist beyond which the

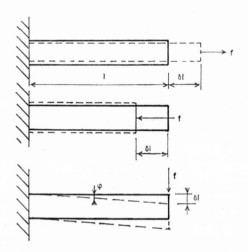

FIG. 6.1. Strain in an elastic material.

relationship ceases to be linear, and the upper limit of stress at which this deviation occurs is known as the limit of proportionality. Stress of lower value than this will result in a strain which will disappear on

TABLE 6.1. Some Properties of Materials

Material	E $(10^6$ lb/in$^2)$	G	σ	Ultimate direct stress $(10^3$ lb/in$^2)$
Steel, mild	29·5	11·5	0·29	61
nickel	28·5	11·0	0·29	115
13% Cr	29·5	11·5	0·28	115
Copper, annealed	15	5·6	0·33	34
Brass, 70/30 ann.	14	5·3	0·33	46
Aluminium, cast	10	3·7	0·34	20–30
Red pine	1·2	0·15		6–9
Oak	1·5	0·15		8–12
Teak	2·4	0·22		8–15

removal of the stress. If stressed beyond the limit of proportionality, permanent deformation of the material may be evident on removal of the stress, although there may be a region of non-linearity without permanent deformation. The ratio *direct stress/direct strain* is known as the modulus of elasticity, E (Young's modulus), and the ratio *shear stress/shear strain* is the modulus of rigidity, or shear modulus, G. Some typical values for commonly used materials are given in Table 6.1.

6.2. RESOLUTION OF STRESS

Consider a section of material, of length AC and unit thickness, which is inclined at an angle of θ to the direction of action of one of

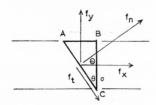

FIG. 6.2. Components of stress.

two mutually perpendicular direct stresses, f_x and f_y (all stresses will be balanced by equal and opposite stresses on the other side of the element). The forces may be resolved normally to the element

$$F_N = F_X \cos \theta + F_Y \sin \theta$$

or

$$f_n \cdot AC \cdot 1 = f_x \cdot BC \cdot 1 \cdot \cos \theta + f_y \cdot AB \cdot 1 \cdot \sin \theta$$

$$f_n = f_x \cdot \cos \theta \cdot BC/AC + f_y \cdot \sin \theta \cdot AB/AC$$

$$= f_x \cdot \cos^2 \theta + f_y \sin^2 \theta \tag{6.1}$$

Resolving tangentially to AC,

$$F_T = F_X \sin \theta - F_Y \cos \theta$$

$$f_t \cdot AC \cdot 1 = f_x \cdot BC \cdot 1 \sin \theta - f_y \cdot AB \cdot 1 \cdot \cos \theta$$

$$f_t = f_x \cdot \sin \theta \cdot BC/AC - f_y \cdot \cos \theta \cdot AB/AC$$

$$= f_x \sin \theta \cdot \cos \theta - f_y \cos \theta \cdot \sin \theta$$

$$= \tfrac{1}{2}(f_x - f_y) \sin 2\theta \tag{6.2}$$

In eqns. (6.1) and (6.2), f_n is the normal stress and f_t the tangential stress on surface AC. It will be seen that, when $\theta = 45°$, there will be a maximum tangential, or shear stress of

$$f_{t(max)} = \tfrac{1}{2}(f_x - f_y) \tag{6.3}$$

It is interesting to note that shear stress cannot exist without a complementary shear stress at right angles to it. Considering the element of material $ABCD$ in Fig. 6.3, and taking moments about point C due to a shear stress q on AB only, would result in an

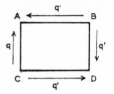

FIG. 6.3. Complementary shear stress.

unbalanced moment. This must therefore be balanced by a force resulting in a shear stress q' on surface AD. Then

$$q'.BD.CD = q.AB.AC, \quad \text{or} \quad q' = q \tag{6.4}$$

6.3. PRINCIPAL STRESSES

When a piece of material is subjected to a system of stresses which are not wholly direct it will always be possible to find planes within the material on which the stresses are wholly direct. From section 6.2 it can be deduced that these planes will be mutually at right angles, since when $\theta = 90°$ there will be no shear stress. In three dimensions there will be three such planes, which are known as principal planes and the stresses on them known as principal stresses. Conversely, the principal stresses may be resolved into the stress system. Logically, one of the principal stresses will be the maximum direct stress in the material; another will be the minimum direct stress in the material. It is often sufficient to consider the two-dimensional case where the minimum principal stress is zero.

In Fig. 6.4 the surface element AC is part of a principal plane on which the principal stress is f, the result of the mutually perpendicular direct stresses f_1 and f_2, and the shear stress q.

Resolving f perpendicular to plane BC,

$$f.AC.\cos\theta = f_1 BC + q.AB$$
$$f.\cos\theta = f_1.BC/AC + q.AB/AC$$
$$= f_1.\cos\theta + q.\sin\theta$$

$$\tan\theta = \frac{f-f_1}{q} \tag{6.5}$$

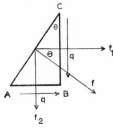

FIG. 6.4. Principal stress.

Resolving f perpendicular to plane AB

$$f.AC.\sin\theta = f_2.AB + q.BC$$
$$f.\sin\theta = f_2.AB/AC + q.BC/AC$$
$$= f_2\sin\theta + q\cos\theta$$

$$\tan\theta = \frac{q}{f-f_2} \tag{6.6}$$

Eliminating θ from eqns. (6.5) and (6.6),

$$\frac{f-f_1}{q} = \frac{q}{f-f_2}$$

$$f^2 - f.f_2 - f.f_1 + f_1 f_2 = q^2$$

$$f^2 - (f_2 + f_1)f + f_1 f_2 - q^2 = 0$$

from which

$$f = \frac{f_2 + f_1}{2} \pm \sqrt{\left[\left(\frac{f_2 + f_1}{2}\right)^2 - f_1 f_2 + q^2\right]}$$

$$= \frac{f_2 + f_1}{2} \pm \sqrt{\left[\left(\frac{f_2 - f_1}{2}\right)^2 + q^2\right]} \tag{6.7}$$

The question now arises as to the meaning of the two solution. Suppose the higher value is taken as being the maximum princip stress f' and the lower value as the principal stress f'' on a plane right angles to the former.

$$\tan \theta' = \frac{f'-f_1}{q} = \frac{\frac{1}{2}(f_2+f_1)+\sqrt{[\frac{1}{4}(f_2-f_1)^2+q^2]}-f_1}{q}$$

$$= \frac{[\frac{1}{2}(f_2-f_1)+\sqrt{(\frac{1}{4}(f_2-f_1)^2+q^2)}][\frac{1}{2}(f_2-f_1)-\sqrt{(\frac{1}{4}(f_2-f_1)^2+q^2)}]}{q[\frac{1}{2}(f_2-f_1)-\sqrt{(\frac{1}{4}(f_2-f_1)^2+q^2)}]}$$

$$= \frac{\frac{1}{4}(f_2-f_1)^2-[\frac{1}{4}(f_2-f_1)^2+q^2]}{q[\frac{1}{2}(f_2-f_1)-\sqrt{(\frac{1}{4}(f_2-f_1)^2+q^2)}]}$$

$$= \frac{-q^2}{q[\frac{1}{2}(f_2+f_1)-\sqrt{(\frac{1}{4}(f_2-f_1)^2+q^2)}-f_1]}$$

that is

$$\tan \theta' = \frac{-q}{f''-f_1} \tag{6.}$$

Equation (6.8) shows that the supposition was correct and tha the second plane lies in the second or fourth quadrant with respe to the first, that is, θ' is displaced by 90° (or 270°) from θ.

The maximum shear stress in the material may now be found fro eqn. (6.3) as

$$q_{max} = \frac{1}{2}(f'-f'')$$
$$= \frac{1}{2}[\frac{1}{2}(f_1+f_2)+\sqrt{(\frac{1}{4}(f_1-f_2)^2+q^2)}-\frac{1}{2}(f_1+f_2)$$
$$+\sqrt{(\frac{1}{4}(f_1-f_2)^2+q^2}$$

or

$$q_{max} = \sqrt{[\frac{1}{4}(f_1-f_2)^2+q^2]} \tag{6.}$$

and

$$f_{max} = \frac{1}{2}(f_1+f_2)+\sqrt{[\frac{1}{4}(f_1-f_2)^2+q^2]} \tag{6.1}$$

6.4. BENDING OF BEAMS AND SHAFTS

Since the lines of action of forces acting on a beam (or a shaft) a rarely collinear with the supports (or bearings), moments resu which cause flexure of the beam. For static conditions there is r overall moment since bending moments caused by loads to one side of a section are balanced by equal and opposite moments due to loads on the other side of the section. Internally, the beam flexure is resisted by stresses in the material.

The bending moment on a beam is found by taking moments of all forces to one side of a section (usually the left-hand side in a diagram), clockwise moments not uncommonly being taken as positive. Figure 6.5 shows a weightless beam (or shaft) carrying a single concentrated load W (such as a fan impeller) acting at a point

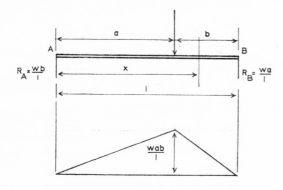

Fig. 6.5. Bending moment under a point load.

which is a distance a from the support A. The beam is said to be simply supported when there is no restraint at the supports, where the beam centre line may take up any slope. The vertical reaction upwards of the support A, R_A, may be found by taking moments about support B, $R_A l - W.b = 0$, from which $R_A = Wb/l$. The reaction at B also acts upwards, and when added to the reaction equals the total load so that there is no net vertical force.

Taking moments about a section distance x from A, where x is less than a, the bending moment will be $R_A.x$. If the distance x is greater than a, the bending moment will be $R_A.x - W(x-a)$. By using special notation the general expression for bending moment at any distance x from A will be

$$M_x = R_A.x - W\{x-a\} \qquad (6.11)$$

where the brackets { } imply that the term within is only valid when x is greater than a. There will be a maximum value of M_x when $x = a$ of $R_A.x = Wab/l$.

A second form of loading worthy of consideration is shown in Fig. 6.6, namely a uniformly distributed load (which includes the weight of the beam itself if this is of uniform cross-section). If the loading per unit length is w, clearly the total weight is wl, and this is shared equally by each support, where $R_A = R_B = wl/2$. At any distance x from the support A, the bending moment due to the

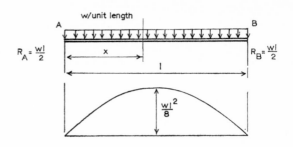

FIG. 6.6. Bending moment under a uniformly distributed load.

support will be $R_A.x$, whilst due to the load wx acting at its centre of gravity distant $x/2$ from A will be $-wx^2/2$, thus

$$M_x = R_A.x - wx.x/2 = wlx/2 - wx^2/2 \qquad (6.12)$$

The distribution of bending moment is parabolic, having a maximum value when $dM_x/dx = 0$, that is, when $wl = 2wx$, or when $x = l/2$, of $wl^2/8$.

Where both point loads and uniformly distributed loads exist together, the moments are additive since the moment due to each is considered separately. Thus the combined effects of the two cases considered above may be derived by adding eqns. (6.11) and (6.12)

$$M_x = Wbx/l - W\{x-a\} + wlx/2 - wx^2/2 \qquad (6.13)$$

Since the point load W gives rise to a discontinuity, it is not quite so easy to find the point of maximum bending moment by differentiation. However, since the moment is effectively the integration of vertical force F, that is, $M = \int F.dx$, the maximum bending moment will occur when the value of F is zero, or when it changes sign in the case of a beam with discontinuous loading.

If the beam is a cantilever, that is a beam with a rigid fixing at one end, the other end being free, it is advisable to take moments from the free end. In Fig. 6.7

$$M_x = W.x + wx^2/2 \qquad (6.14)$$

and the maximum value will be at the support where the value of x is maximum.

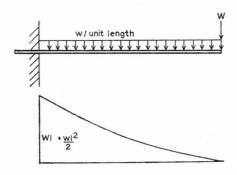

FIG. 6.7. Bending moment of a cantilever.

A simple method of drawing a bending moment diagram is shown in Fig. 6.8. Using Bow's notation, the spaces between the forces on the beam are denoted by A, B, C and D, the scale of the diagram being S_1. A diagram of forces is drawn to scale S_2, with length bc representing W_1 (the force between spaces B and C), cd representing W_2, and da and ab representing reactions R_B and R_A respectively. Since all the forces act in a vertical direction, the diagram $abcd$ is a straight line. A pole O is chosen anywhere and is joined by straight lines to points a, b, c and d. Then to the same scale of length, S_1, as the original loading diagram, line PS is drawn between R_A and W_1, parallel to Ob; lines SV, VX and XP are drawn between the lines of action of the appropriate loads and parallel to lines Oc, Od and Oa respectively.

M

Now triangles PQS and PRS are similar to triangles Oba and ObH, the line OH being horizontal. Thus

$$QS/PR = ab/OH$$

and

$$QS = ab . PR/OH$$

Now $PR = l_1/S_1$, and $ab = R_A/S_1$, hence

$$QS = R_A l_1/S_1 S_2 OH$$

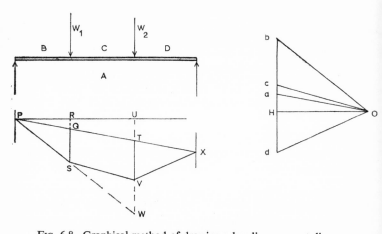

Fig. 6.8. Graphical method of drawing a bending moment diagram.

But the bending moment under the load W_1, $M_1 = R_A l_1$, or

$$M_1 = QS . S_1 S_2 OH \tag{6.15}$$

Also, triangles PTW, PUW and SVW are similar to triangles Oba, ObH and Ocb respectively, and

$$TW = ab . PU/OH$$

and

$$VW = bc . RU/OH$$

$$TV = TW - VW = ab . PU/OH - bc . RU/OH$$

Now $PU = l_2/S_1$; $bc = W_1/S_2$; $RU = (l_2 - l_1)/S_1$,

$$TV . OH = R_A l_2/S_2 S_1 - W_1(l_2 - l_1)/S_2 S_1$$

But the bending moment under W_2, $M_2 = R_A l_2 - W_1(l_2 - l_1)$, or

$$M_2 = TV . S_1 S_2 OH \tag{6.16}$$

It is probable that this method of drawing the bending moment has no advantage over the use of an equation such as eqn. (6.13) for point loads or uniformly distributed loads, but where there is a non-uniformly distributed load, the rate of loading may be plotted to scale as shown in Fig. 6.9. The area of this diagram represents the total load, which may be split up into a number of segments, each representing part of the load. Each part load may be taken as a point load acting at the centre of area of the segment. From these, a

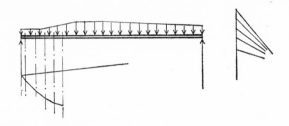

FIG. 6.9. Application of graphical method to a non-uniformly
distributed load.

polar diagram may be drawn. The degree of accuracy depends on the number of segments chosen.

6.5. STRESSES RESULTING FROM SIMPLE BENDING

Where a beam is subjected to a simple bending moment (that is, in the absence of vertical force, or *shear force* as it is known) it is possible, by making certain simple assumptions, to deduce the stress set up in the material. It is found that these relationships can be applied with sufficient accuracy to other cases, although the maximum stress is generally likely to occur when the bending moment is maximum, that is, when the shear force is zero.

The assumptions made are:

(i) the material of the beam is homogeneous and elastic,
(ii) the modulus of elasticity has the same value in compression as in tension,
(iii) transverse fibres remain at right angles to the plane of bending at all times, and

(iv) an axis exists which is the same length before and after bending (that is, it suffers no strain), and is known as the *neutral axis*.

Consider, in Fig. 6.10, a longitudinal plane of original length AB to undergo bending and to be strained to $A'B'$, having a radius

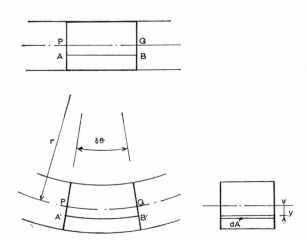

FIG. 6.10. Beam under simple bending.

$r+y$. The radius of the neutral axis is r, and the angle subtended to the centre of curvature is $\delta\theta$. Then the strain on the section is

$$e = \frac{A'B' - PQ}{PQ} = \frac{(r+y).\delta\theta - r.\delta\theta}{r.\delta\theta} = \frac{y}{r}$$

But also

$$e = \frac{f}{E}$$

Thus

$$\frac{f}{E} = \frac{y}{r} \quad \text{or} \quad f = \frac{Ey}{r} \tag{6.17}$$

The bending stress is proportional to the distance from the neutral axis, and the maximum stress will thus occur in the material fibres

arthest from this axis. To one side of the neutral axis the stress will
be tensile, whilst to the other side it will be compressive.

Since there is no strain on the neutral axis, there will also be no
stress. Thus, with respect to the neutral axis,

$$\int f.dA = 0$$

and since $f = Ey/r$,

$$\int \frac{Ey}{r}.dA = 0$$

$$\frac{E}{r}\int y.dA = 0$$

Since $\int y.dA = 0$ at the centre of area of the section, the neutral
axis must pass through this point.

Bending of the beam is resisted by the internal stresses which set
up an equal and opposite resisting moment of

$$M = \int f.dA.y$$

$$= \int \frac{E}{r}.y^2dA$$

$$= \frac{E}{r}\int y^2dA = \frac{EI}{r}$$

where $\int y^2dA = I$, known as the second moment of area of the
section. Thus

$$\frac{M}{I} = \frac{E}{r} = \frac{f}{y} \qquad (6.18)$$

The value of I/y, where y is the distance from the neutral axis to
the farthest material fibre, is known as the section modulus Z. Thus
the maximum bending stress

$$f_{max} = M/Z \qquad (6.19)$$

Standard rolled steel sections, such as joists, tees, angles and channels,
will be found to have their properties tabulated.[21] These include
position of the neutral axis in a number of planes, second moment of
area and section modulus. It is a simple matter to calculate these

values for rectangular sections. The neutral axis will pass through the section at half the depth and

$$I = \int_{-d/2}^{+d/2} b \cdot y^2 \mathrm{d}y = \frac{bd^3}{12} \qquad (6.20)$$

where b is the breadth, and d the depth, of the section.

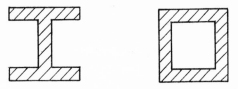

FIG. 6.11. Cut-away sections.

Other symmetrical sections such as those shown in Fig. 6.11 may be dealt with by subtracting I for the spaces from the value of I for the enveloping shape about a common neutral axis.

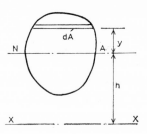

FIG. 6.12. Parallel axes theorem.

For more complex shapes, two simple theorems are useful. The parallel axes theorem, illustrated by Fig. 6.12, gives the second moment of area about any axis parallel to the neutral axis NA for

$$I_{xx} = \int (h+y)^2 \mathrm{d}A$$

$$= \int h^2 \mathrm{d}A + \int 2yh \cdot \mathrm{d}A + \int y^2 \mathrm{d}A$$

But $y \cdot \mathrm{d}A = 0$ since the neutral axis passes through the section centroid, y being the distance from the centroid to the element of area, thus

$$I_{xx} = \int h^2 dA + y^2 dA$$
$$= A \cdot h^2 + I_{NA} \tag{6.21}$$

The perpendicular axes theorem is illustrated in Fig. 6.13. The second moment of an element of area dA about an axis ZZ, perpendicular to the plane of the diagram (the polar second moment of area, J) may be written

$$I_{zz} = J = \int r^2 dA$$
$$= \int (x^2 + y^2) dA$$
$$= \int x^2 dA + \int y^2 dA = I_{yy} + I_{xx} \tag{6.22}$$

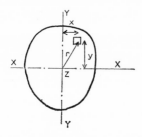

FIG. 6.13. Perpendicular axes theorem.

This second theorem is particularly useful for finding the second moment of area about a diameter of a circular section.

$$J = \int_0^r 2\pi r^3 \cdot dr = \pi r^4 / 2 = \pi d^4 / 32$$

Since the second moment of area is the same about any diameter,

$$I_{xx} = I_{yy} = J/2 = \pi d^4 / 64 \tag{6.23}$$

Sometimes a beam or shaft may be subjected to an end loading in addition to the loads causing bending. In such a case, if the end load causes tension, there will be a stress equal to the load divided by the cross-sectional area. The combined stress f will be the algebraic sum of the bending tensile stress f_b and the direct stress f_t, that is,

$$f = f_b + f_t \tag{6.24}$$

This has the effect of increasing the stress to one side of the neutral axis, and decreasing it to the other. If the end load causes a compressive stress, this has the effect of reversing the signs in eqn. (6.24) where appropriate.

Although the simple theory for bending stress outlined above has been considered primarily for a shaft or beam, it may also be used to give an approximation to the stress in a rectangular flat plate subjected to a uniform pressure and supported at the perimeter. The plate is considered to have a probable maximum bending moment (and a maximum bending stress) on a diagonal section such as AC in Fig. 6.14. The reactions on the sides AB and BC, of lengths a and

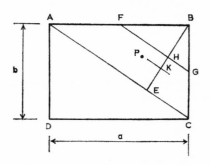

FIG. 6.14. Bending of a rectangular plate supported at the edges.

b respectively, will effectively be at the mid-points F and G, taken as acting along the line FG. The value of the reaction R will be half the total force due to the pressure difference p across the plate.

$$R = \tfrac{1}{2} p . a . b \qquad (6.25)$$

Taking moments about AC,

$$M = R . HE - \tfrac{1}{2} p . a . b . KE \qquad (6.26)$$

where KE is the perpendicular distance from the centre of pressure P to the diagonal AC. Inserting the value of R from eqn. (6.25) in eqn. (6.26),

$$M = \tfrac{1}{2} p . a . b . BE/2 - \tfrac{1}{2} p . a . b . BE/3$$

$$= p . a . b . BE/12$$

But $BE/BC = AB/AC$, or $BE = BC.AB/AC = a.b/\sqrt{(a^2+b^2)}$
and

$$M = \frac{p.a^2b^2}{12\sqrt{(a^2+b^2)}}$$

Now

$$f = My/I = \frac{M.t.12}{2AC.t^3} = \frac{6M}{t^2\sqrt{(a^2+b^2)}}$$

$$= \frac{p.a^2b^2}{2t^2(a^2+b^2)} \tag{6.27}$$

6.6. SHEAR STRESS DUE TO TORSION

Power is transmitted by a shaft by torque and speed of rotation. Considering a force F at the periphery of a shaft, the work done per revolution will be $F \times 2\pi r$. If there are n revolutions per unit time, the power transmitted will be $2\pi.n.F.r$, or $2\pi.n.T$, where T is the torque, or twisting moment, $F \times r$. If the torque is in lbf ft and n is in revolutions per minute,

$$\text{horsepower} = 2\pi.n.T/33{,}000 \tag{6.28}$$

If the torque is in kgf.m and n is in revolutions per minute,

$$\text{power} = 2\pi.n.T/6120 \text{ kW} = 2\pi n.T/4500 \text{ metric h.p.} \tag{6.29}$$

The torque transmitted by a shaft results in strain, and a stress which resists the torque. Considering a shaft such as shown in Fig. 6.15, a line AB on the unstrained shaft will become $A'B$ when under torsion, and results in a shear strain,

$$\phi = \frac{AA'}{AB} = \frac{r_0\theta}{l} = \frac{q_0}{G}$$

Thus shear stress at radius r_0,

$$q_0 = \frac{Gr_0\theta}{l} \tag{6.30}$$

It follows that the shear stress at any radius r,

$$q = \frac{Gr\theta}{l} \tag{6.31}$$

and

$$\frac{q}{q_0} = \frac{r}{r_0} \tag{6.32}$$

Equation (6.32) indicates that the shear stress resisting torsion is proportional to the radius, and is maximum at the outer fibres of the

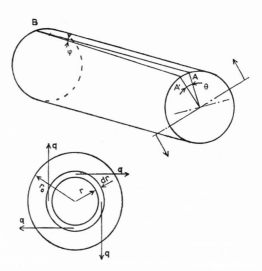

Fig. 6.15. Torsion of a shaft.

material. Considering a radial element of a circular shaft, the resisting torque will be

$$dT = q \cdot 2\pi r \cdot dr \cdot r = \frac{q_0}{r_0} \cdot 2\pi r^3 dr$$

(i) For a solid shaft:

$$T = \frac{2\pi q_0}{r_0} \int_0^{r_0} r^3 dr = \frac{\pi q_0 r_0^4}{2r_0} = \frac{\pi q_0 d^3}{16} \tag{6.33}$$

(ii) For a hollow shaft:

$$T = \frac{2\pi q_0}{r} \int_{r_1}^{r_0} r^3 \, dr = \frac{\pi q_0}{2r_0}(r_0^4 - r_1^4)$$

$$= \frac{\pi q_0}{16 d_0}(d_0^4 - d_1^4) \tag{6.34}$$

where d_1 is the inner diameter. Since the polar second moment of area, J, in case (i) is $\pi d_0^4/32$, and in case (ii) is $\pi(d_0^4 - d_1^4)/32$, eqns. (6.33) and (6.34) may be written

$$T = \frac{J}{r_0} \cdot q_0 \tag{6.35}$$

Equation (6.35) should not, however, be used as a general expression to find the shear stress in shafts of any shape of cross-section.

6.7. STRESSES DUE TO COMBINED BENDING AND TORSION

In a shaft subjected to both bending and torsion, there will be a direct stress due to bending of $f_0 = My_0/I$, and a shear stress due to torsion of $q_0 = Tr_0/J$. From eqn. (6.10), the maximum direct stress will be

$$f_{max} = \tfrac{1}{2}f_0 + \sqrt{(\tfrac{1}{4}f_0^2 + q_0^2)} \tag{6.36}$$

For a solid circular shaft, $I = \pi d_0^4/64$, $y_0 = d_0/2$ and $J = \pi d_0^4/32$, which, on substitution in eqn. (6.36) gives

$$f_{max} = \frac{16}{\pi d_0^3}[M + \sqrt{(M^2 + T^2)}] \tag{6.37}$$

Similarly, the maximum shear stress from eqn. (6.9) will be

$$q_{max} = \sqrt{(\tfrac{1}{4}f_0^2 + q_0^2)}$$

$$= \frac{16}{\pi d_0^3}\sqrt{(M^2 + T^2)} \tag{6.38}$$

Thus a shaft should be designed such that the diameter d_0 allows both f_{max} and q_{max} to be within permissible limits (about 18,000 lb/in^2 (1250 kg/cm^2) and 12,000 lb/in^2 (840 kg/cm^2) respectively, for steel).

6.8. DEFLECTION OF SHAFTS AND BEAMS

Figure 6.16 shows the centre line of a beam which has been subjected to simple bending, the effect being considerably exaggerated for clarity. An element AB, of length ds, subtends an angle of di to some point to which the radius is r.

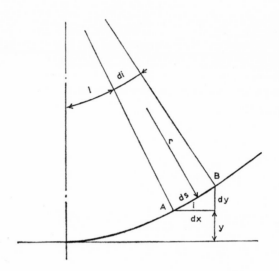

FIG. 6.16. Slope and deflection of a bent beam.

Then

$$r \cdot di = ds \quad \text{or} \quad \frac{di}{ds} = \frac{1}{r}$$

But the gradient (or slope) of the element, $\tan i = dy/dx$, and if i is very small, $i = dy/dx$, and $ds = dx$ very nearly.

Thus

$$\frac{1}{r} = \frac{d}{ds}\left(\frac{dy}{dx}\right) = \frac{d^2y}{dx^2}$$

Also, $1/r = M/EI$, thus

$$\frac{d^2y}{dx^2} = \frac{M}{EI} \tag{6.39}$$

rom this

$$i = \frac{dy}{dx} = \int \frac{M}{EI} \cdot dx + \text{const} \tag{6.40}$$

nd

$$y = \int \left[\int \frac{M}{EI} \cdot dx + \text{const}_1 \right] \cdot dx + \text{const}_2 \tag{6.41}$$

here y is the vertical displacement, or deflection, of the beam
:ntre line.

For a beam simply supported at its ends, having a length between
apports of l, and a single point load W at a distance of a from one
apport (as in Fig. 6.5),

$$M_x = Wbx/l - W\{x-a\}$$

onsidering E and I as constant, and integrating the bracketed term
; a whole,

$$i_x = \frac{1}{EI} \int M_x dx = \frac{W}{EI} \left[\frac{bx^2}{2l} - \tfrac{1}{2}\{x-a\}^2 + C \right]$$

$$y_x = \frac{1}{EI} \int i_x dx = \frac{W}{EI} \left[\frac{bx^3}{6l} - \tfrac{1}{6}\{x-a\}^3 + Cx + D \right]$$

here C and D are constants of integration which can be evaluated
om a knowledge of the beam end conditions. For example, in this
ise, $y = 0$ at $x = 0$ (at the left-hand support), thus $D = 0$. Also,
= 0 at $x = l$ (at the right-hand support), from which

$$0 = \frac{bl^2}{6} - \frac{b^3}{6} + Cl$$

ving

$$C = \frac{-b}{6l}(l^2 - b^2)$$

hus

$$y_x = \frac{W}{EI} \left[\frac{bx^3}{6l} - \tfrac{1}{6}\{x-a\}^3 - \frac{bx}{6l}(l^2 - b^2) \right] \tag{6.42}$$

Often the deflection of the beam under the load is of interest, tha
is, when $x = a$:

$$y_a = \frac{W}{EI}\left[\frac{ba^3}{6l} - \frac{ba}{6l}(l^2 - b^2)\right]$$

$$= \frac{Wab}{6EIl}[a^2 - (a+b)^2 + b^2]$$

$$= \frac{Wab}{6EIl}[a^2 - a^2 - 2ab - b^2 + b^2]$$

$$= -\frac{Wa^2b^2}{3EIl} \tag{6.4}$$

The negative sign results from y being measured towards a hor
zontal tangent whereas, in eqn. (6.41), y has been measured awa
from the tangent. However, it could be argued that the sign
arbitrary since the positive direction of the bending moment cou
equally refer to an anticlockwise moment. If the maximum deflectic
is required, the position at which it occurs must first be determine
Since $i = dy/dx$, equating this to zero and finding the value of x w
give the correct position, although in view of the special brackete
term some assumption must be made as to the approximate positic
at which this will occur.

For a beam with a uniformly distributed load, the bendi
moment at any distance x from one support is given by

$$M_x = wlx/2 - wx^2/2$$

$$i_x = \frac{w}{EI}\left[\frac{lx^2}{4} - \frac{x^3}{6} + C\right]$$

and

$$y_x = \frac{w}{EI}\left[\frac{lx^3}{12} - \frac{x^4}{24} + Cx + D\right]$$

Since $y = 0$ at $x = 0$, $D = 0$, and $y = 0$ at $x = l$,

$$0 = \frac{l^4}{12} - \frac{l^4}{24} + Cl$$

Thus

$$C = -\frac{l^3}{24}$$

nd

$$y_x = \frac{w}{EI}\left[\frac{lx^3}{12} - \frac{x^4}{24} - \frac{l^3x}{24}\right] \qquad (6.44)$$

The deflection of the greatest interest in this case is probably that
t the mid-point (which is also the maximum deflection for this form
f loading) where $x = l/2$:

$$y_{\max} = \frac{w}{EI}\left[\frac{l^4}{96} - \frac{l^4}{384} - \frac{l^4}{48}\right]$$

$$= \frac{wl^4}{384EI}[4 - 1 - 8]$$

$$y_{\max} = -\frac{5wl^4}{384EI} \qquad (6.45)$$

Where the above cases occur together, it can be seen that the
deflection at any point may be found by adding together eqns. (6.42)

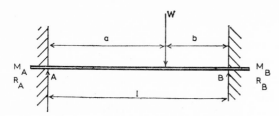

Fig. 6.17. Beam with restrained fixings.

nd (6.44). The maximum deflection will occur where the slope,
= 0.

Where the ends of the beam are restrained from rotating as shown
n Fig. 6.17, there will be restraining moments of unknown value at
he fixing points, and the reactions cannot be calculated. Thus at
ny distance x from one support,

$$M_x = M_A + R_A x - W\{x - a\}$$

$$EIi_x = \int M_x dx = M_A x + \frac{R_A x^2}{2} - \frac{W}{2}\{x - a\}^2 + C$$

$$EIy_x = \int i_x dx = \frac{M_A x^2}{2} + \frac{R_A x^3}{6} - \frac{W}{6}\{x - a\}^2 + Cx + D$$

Now, $y = 0$ at $x = 0$, thus $D = 0$.

Also, $i = 0$ at $x = 0$, thus $C = 0$.

In addition, y and i are both zero when $x = l$, giving

$$0 = M_A l + \frac{R_A l^2}{2} - \frac{Wb^2}{2}$$ (6.46)

and

$$0 = \frac{M_A l^2}{2} + \frac{R_A l^3}{6} - \frac{Wb^3}{6}$$ (6.47)

Multiplying eqn. (6.47) by $2/l$ gives

$$0 = M_A l + \frac{R_A l^2}{3} - \frac{Wb^3}{3l}$$

and, on subtracting from eqn. (6.46),

$$0 = \frac{R_A l^2}{6} - W \left[\frac{b^2}{2} - \frac{b^3}{3l} \right]$$

and

$$R_A = \frac{6W}{l^2} \left[\frac{b^2}{2} - \frac{b^3}{3l} \right] = \frac{Wb^2}{l^3} [3l - 2b]$$

$$= \frac{Wb^2}{l^3} (3a + b)$$

On substitution back in eqn. (6.46), this gives

$$M_A = -\frac{Wb^2 l}{2l^3}(3a + b) + \frac{Wb^2}{2l} = -\frac{Wb^2}{2l^2}(3a + b - l)$$

$$= -\frac{Wab^2}{l^2}$$

and the general expression for deflection becomes

$$y_x = \frac{1}{EI} \left[\frac{Wb^2 x^3}{6l^3}(3a + b) - \frac{Wab^2 x^2}{2l^2} - \frac{W}{6} \{x - a^3\} \right]$$ (6.48)

The deflection under the load at $x = a$ becomes

$$y_a = \frac{1}{EI}\left[\frac{Wb^2a^3}{6l^3}(3a+b)-\frac{Wa^3b^2}{2l^2}\right] = \frac{Wa^3b^2}{6EIl^3}(3a+b-3l)$$

$$= \frac{Wa^3b^2}{6EIl^3}(3a+b-3a-3b)$$

$$y_a = -\frac{Wa^3b^3}{3EIl^3} \tag{6.49}$$

For a uniformly distributed load,

$$M_x = M_A+R_Ax-\frac{wx^2}{2} \tag{6.50}$$

$$EIi_x = M_x\,dx = M_Ax+\frac{R_Ax^2}{2}-\frac{wx^3}{6}+C$$

and

$$EIy_x = i_x\,dx = \frac{M_Ax^2}{2}+\frac{R_Ax^3}{6}-\frac{wx^4}{24}+Cx+D$$

Since both i_x and y_x are zero when $x = 0$, C and D are zero. Also, i_x and y_x are zero when $x = l$, giving

$$0 = M_Al+\frac{R_Al^2}{2}-\frac{wl^3}{6} \tag{6.51}$$

$$0 = \frac{M_Al^2}{2}+\frac{R_Al^3}{6}-\frac{wl^4}{24} \tag{6.52}$$

Multiplying eqn. (6.52) by $2/l$ and subtracting from eqn. (6.51) gives

$$0 = \frac{R_Al^2}{6}-\frac{wl^3}{12} \quad \text{or} \quad R_A = \frac{wl}{2}$$

which is to be expected, since the beam is symmetrical. On substituting this value into eqn. (6.51),

$$0 = M_Al+\frac{wl^3}{4}-\frac{wl^3}{6}$$

from which

$$M_A = -\frac{wl^2}{12} \tag{6.53}$$

By substituting this value back into eqn. (6.50), the maximum bending moment is found to have the same value, $wl^2/12$.

The deflection

$$y_x = \frac{1}{EI}\left[\frac{wlx^3}{12} - \frac{wl^2x^2}{24} - \frac{wx^4}{24}\right] \qquad (6.54)$$

The deflection at mid-span, when $x = \frac{1}{2}l$ (the maximum deflection),

$$y_{max} = \frac{1}{EI}\left[\frac{wl^4}{96} - \frac{wl^4}{96} - \frac{wl^4}{384}\right]$$

$$= -\frac{wl^4}{384EI} \qquad (6.55)$$

6.9. A GRAPHICAL METHOD OF FINDING THE DEFLECTION OF A BEAM

The graphical method of finding the bending moment of a beam, described in section 6.4, may be extended to find the deflection. This method is particularly useful for stepped shafts where the second moment of area varies along the length. An M/I diagram is considered in the same light as a diagram of rate of distributed loading. This is subdivided into a number of sections, the " area " of each section being considered (by analogy to the loading on a section of a beam) to act at the centroid. A vertical diagram of " forces " is drawn for the " areas " and a polar diagram constructed. In the same way as for the bending moment, lines parallel to the appropriate rays are drawn in the corresponding spaces between centroids (for example, the line in space A is parallel to ray oa). A vertical measurement Y on the resulting diagram gives the deflection y at that point along the beam, since (as eqns. (6.15) and (6.16))

$$y = \frac{YS_1S_3OK}{E} \qquad (6.56)$$

where S_1 is the scale of length in terms of length of beam per length of diagram, S_3 is the scale of area used on the diagram of " forces " in terms of the units of Ml/I (force per length^2) per unit area of M/I diagram, OK is the normal distance to the pole on the diagram of " forces ", and E is the modulus of elasticity of the material.

The resulting deflection diagram is a series of straight lines rather than a smooth curve. The closeness of the approximation depends on the number of subdivisions of the M/I diagram.

Examples are given in Fig. 6.18 and at the end of the chapter.

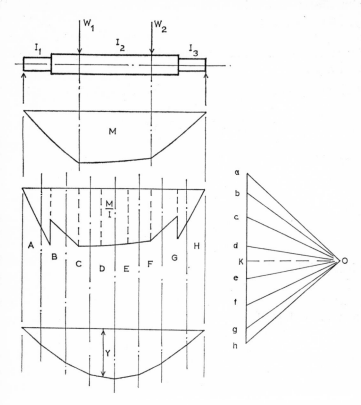

FIG. 6.18. Graphical method for the deflection of a beam or shaft.

6.10. TRANSVERSE VIBRATIONS OF BEAMS

Equation (4.12) shows that the natural frequency of vibration of a weight on a spring is related only to the static deflection of the weight. The expressions derived in section 6.8 show that the deflection of the load on a beam is proportional to the load itself. Thus a beam is a spring, and the natural frequency of vibration of the load, f_0, will be given by eqn. (4.12) as

$$f_0 = \frac{1}{2\pi} \sqrt{\frac{g}{\delta}},$$

where y is the deflection under the load.

The same result may be obtained by assuming that the load moves with simple harmonic motion, and equating the energy in the system at the extreme position of motion (where there is no kinetic energy) to that at the mean position (where there is no strain energy). Suppose the amplitude of vibration to be a, and the angular frequency to be ω_0, then

strain energy at extreme position = kinetic energy at mean position

$$\int_0^a ka \cdot da = \tfrac{1}{2}ka^2 = \frac{1}{2}\frac{W}{g}(a\omega_0)^2 \qquad (6.57)$$

from which

$$\omega_0 = \sqrt{\frac{kg}{W}} = \sqrt{\frac{g}{y}}$$

and

$$f_0 = \frac{\omega_0}{2\pi} = \frac{1}{2\pi}\sqrt{\frac{g}{y}} = 3\cdot13\sqrt{\frac{1}{y''}}\ \text{c/s}$$

where y'' is in inches.

Equation (6.57) may be rewritten in terms of the beam deflection under the load, y, since $a = (a/y) \times y$, and the strain energy resulting from a deflection of a would be

$$\tfrac{1}{2}k\left(\frac{a}{y}\cdot y\right)^2 = \tfrac{1}{2}Wy\left(\frac{a}{y}\right)^2, \quad \text{since } W = ky.$$

Thus

$$\tfrac{1}{2}Wy\left(\frac{a}{y}\right)^2 = \frac{1}{2}\frac{W}{g}\left(\frac{a}{y}\cdot y\omega_0\right)^2$$

for

$$\tfrac{1}{2}Wy = \frac{1}{2}\cdot\frac{W}{g}(y\omega_0)^2 \qquad (6.58)$$

This gives the same result for ω_0 as before. If there are a number of point loads on a beam, eqn. (6.58) may be extended to cover the case where all the loads vibrate in phase:

$$\sum\tfrac{1}{2}Wy = \sum\frac{1}{2}\cdot\frac{W}{g}(y\omega_0)^2$$

from which

$$\omega_0^2 = \frac{g\Sigma Wy}{\Sigma Wy^2} \qquad (6.59)$$

If the reasonable assumption is made that a uniformly loaded beam will vibrate with a shape similar to that of the shape under a static deflection, eqn. (6.59) may itself be extended to:

$$\omega_0^2 = \frac{g\int_0^l y\,.\,\mathrm{d}y}{\int_0^l y^2\mathrm{d}y}$$

and substituting in this equation the value of y from eqn. (6.44),

$$\omega_0^2 = \frac{24EIg\,.\int_0^l(2lx^3 - x^4 - l^3x)}{w\int_0^l(x^8 - 4lx^7 + 4l^2x^6 + 2l^3x^5 - 4l^4x^4 + l^6x^2)}$$

$$= \frac{24EIg\,.\left[\dfrac{2lx^4}{4} - \dfrac{x^5}{5} - \dfrac{l^3x^2}{2}\right]_0^l}{w\left[\dfrac{x^9}{9} - \dfrac{4lx^8}{8} + \dfrac{4l^2x^7}{7} + \dfrac{2l^3x^6}{6} - \dfrac{4l^4x^5}{5} + \dfrac{l^6x^3}{3}\right]_0^l}$$

$$= \frac{24EIg\left[\dfrac{l^5}{2} - \dfrac{l^5}{5} - \dfrac{l^5}{2}\right]}{w\left[\dfrac{l^9}{9} - \dfrac{l^9}{2} + \dfrac{4l^9}{7} - \dfrac{4l^9}{5} + \dfrac{2l^9}{3}\right]}$$

$$= \frac{24EIg}{wl^4}\,.\,\frac{630}{155} = \frac{97\cdot6EIg}{wl^4}$$

and, since from eqn. (6.45),

$$y_{\max} = \frac{5wl^4}{384EI}$$

$$\omega_0^2 = \frac{1\cdot27g}{y_{\max}}$$

and

$$f_0 = \frac{1}{2\pi}\sqrt{\left(\frac{1\cdot27g}{y_{\max}}\right)}$$

$$= 3\cdot53\sqrt{\left(\frac{1}{y''_{\max}}\right)}\text{ c/s,}\qquad(6.60)$$

where $y''_{\max}$ is in inches.

This frequency is actually only the lowest of a number of possible frequencies of natural vibration. It is known as the *fundamental* frequency, and is often of the greatest interest.

If a beam of appreciable weight has in addition a number of point loads, it is possible to estimate the overall natural frequency of vibration using an empirical relationship:[22]

$$\frac{1}{f_0^2} = \frac{1}{f_{01}^2} + \frac{1}{f_{02}^2} + \dots + \frac{1}{f_{0n}^2} + \dots + \frac{1}{f_{0w}^2} \qquad (6.61)$$

where f_{01}, f_{02}, etc., are the frequencies of natural vibration of the point loads on the beam taken separately and in the absence of any other load, and f_{0w} is the natural frequency of the beam alone as given by eqn. (6.60).

6.11. CRITICAL SPEEDS OF ROTATING SHAFT SYSTEMS

Since the effect of gravity is irrelevant to the argument, it is convenient to consider a vertical shaft on which is mounted a heavy rotor, whose centre of gravity is displaced (due to imperfect manufacture) a distance δ from the axis of rotation. Due to centrifugal force, the shaft will bend when being rotated, this bending being resisted by the shaft stiffness, thus

$$k \cdot x = \frac{W}{g} \cdot \omega^2(x+\delta)$$

since

$$\omega_0^2 = \frac{kg}{w}$$

$$\omega_0^2 x = \omega^2 x + \omega^2 \delta$$

$$x = \frac{\omega^2 \delta}{\omega_0^2 - \omega^2} \qquad (6.62)$$

From eqn. (6.62), it can be seen that when $\omega_0 < \omega$, the centre of gravity of the rotor will be displaced by an amount $\delta - x$ from the axis of rotation, since x is negative. When $\omega_0 = \omega$, x tends to become infinite with resultant failure of the shaft. Such speed, known as the critical speed n_c, must be avoided at all cost. That is to say, the critical speed in revolutions per unit time is the same as the natural frequency of transverse vibration of the shaft system in cycles per unit time. It is customary for fan shafts to run at speeds not exceeding two-thirds of the lowest critical speed (corresponding to the lowest frequency of natural vibration).

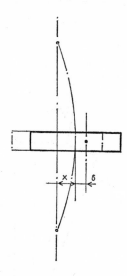

FIG. 6.19. Bending of a vertical shaft under rotation.

6.12. STRESSES IN FAN BACKPLATES AND SHROUDS

The complete fan impeller is a complex unit for which to assess the stresses. For simplicity, stresses in shroud and backplate are considered in isolation from those in the blades.

Considering a disc of homogeneous material of weight per unit volume w, whose uniform thickness t is small compared with the diameter, an element (as shown in Fig. 6.20) will have a volume of

$$[\tfrac{1}{2}(r+\delta r)^2\delta\theta - \tfrac{1}{2}r^2\delta\theta]t = r.\delta r.\delta\theta.t$$

neglecting second order small quantities. When this disc rotates with an angular velocity ω about the axis, there will be direct stresses tending to strain the disc in a radial direction f_r and in a tangential direction f_h. The centrifugal force on the element will be

$(w/g).r.\delta r.\delta\theta.t.\omega^2 r$. This force will be resisted by the stresses on the section acting in a radial direction resulting in a force of

$$(f_r+\delta f_r)(r+\delta r)t.\delta\theta-f_r r.t.\delta\theta-2f_h\delta r.t.\sin(\delta\theta/2)$$

$$=(f_r r+f_r\delta r+r.\delta f_r+\delta f_r\delta r-f_r r-f_h\delta r)t.\delta\theta$$

(since $\sin\delta\theta$ is very nearly equal to $\delta\theta$)

$$=(f_r\delta r+r.\delta f_r-f_h\delta r)t.\delta\theta$$

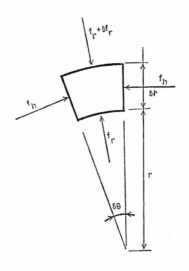

FIG. 6.20. Element of rotating material.

and since there is no net force,

$$\frac{w}{g}.\omega^2 r^2\delta r.\delta\theta.t+(f_r\delta r+r.\delta f_r-f_h\delta r)t.\delta\theta=0$$

which, in the limit, as δr approaches zero, becomes

$$\frac{w}{g}.\omega^2 r^2+\frac{\mathrm{d}}{\mathrm{d}r}(r.f_r)-f_h=0 \qquad (6.63)$$

Considering the strain on the element due to stresses f_r and f_h the radius increases by u:

(a) circumferential strain,

$$e_h = \frac{2\pi(r+u)-2\pi r}{2\pi r} = \frac{u}{r}$$

also

$$e_h = \frac{u}{r} = \frac{1}{E}(f_h - \sigma f_r) \tag{6.64}$$

(b) The element width becomes

$$r + \delta r + u + \delta u - (r+u) = \delta r + \delta u$$

radial strain,

$$e_r = \frac{\delta r + \delta u - \delta r}{\delta r} = \frac{\delta u}{\delta r} = \frac{du}{dr} \quad \text{(in the limit)}$$

also

$$e_r = \frac{du}{dr} = \frac{1}{E}(f_r - \sigma f_h) \tag{6.65}$$

Multiplying eqn. (6.64) by σ and adding to eqn. (6.65) gives

$$f_r = \frac{E}{1-\sigma^2}\left(\sigma \frac{u}{r} + \frac{du}{dr}\right) \tag{6.66}$$

and, multiplying eqn. (6.65) by σ and adding to eqn. (6.64) gives

$$f_h = \frac{E}{1-\sigma^2}\left(\frac{u}{r} + \sigma \frac{du}{dr}\right) \tag{6.67}$$

Substituting eqns. (6.66) and (6.67) in eqn. (6.63),

$$\frac{w}{g} \cdot \omega^2 r^2 + \frac{d}{dr}\left[\frac{E}{1-\sigma^2}\left(\sigma u + r \cdot \frac{du}{dr}\right)\right] - \frac{E}{1-\sigma^2}\left(\frac{u}{r} + \sigma \frac{du}{dr}\right) = 0$$

$$\frac{(1-\sigma^2) \cdot w \cdot \omega^2 r^2}{Eg} + \sigma \frac{du}{dr} + \frac{du}{dr} + r \frac{d^2u}{dr^2} - \frac{u}{r} - \sigma \frac{du}{dr} = 0$$

giving

$$\frac{d^2u}{dr^2} + \frac{d}{dr}\left(\frac{u}{r}\right) = -\frac{w \cdot \omega^2 r(1-\sigma^2)}{gE} \tag{6.68}$$

$$\left(\text{since } \frac{d}{dr}\left(\frac{u}{r}\right) = \frac{1}{r} \cdot \frac{du}{dr} - \frac{u}{r^2}\right)$$

The complementary function of the solution of eqn. (6.68) can be found by equating the right-hand side to zero and integrating the left-hand side:

$$\frac{du}{dr} + \frac{u}{r} = 2A \qquad (6.69)$$

where A is a constant.

$$r.\frac{du}{dr} + u = \frac{d}{dr}(ur) = 2Ar$$

and integrating again

$$ur = Ar^2 + B$$

where B is a second constant, or

$$\frac{u}{r} = A + \frac{B}{r^2} \qquad (6.70)$$

Substituting eqn. (6.70) in eqn. (6.69) gives

$$\frac{du}{dr} + A + \frac{B}{r^2} = 2A$$

from which

$$\frac{du}{dr} = A - \frac{B}{r^2} \qquad (6.71)$$

To find the particular integral, it may be assumed that $u = Cr^3$, thus, substituting this in eqn. (6.68),

$$6Cr + 3Cr - Cr = -\frac{w.\omega^2 r(1-\sigma^2)}{gE}$$

or

$$C = -\frac{w.\omega^2(1-\sigma^2)}{8gE} \qquad (6.72)$$

Now, $u/r = Cr^2$, and adding this to eqn. (6.70) gives the complete solution to eqn. (6.68), thus

$$\frac{u}{r} = A + \frac{B}{r^2} + Cr^2 \qquad (6.73)$$

Also, adding $du/dr = 3Cr^2$ to eqn. (6.71) gives

$$\frac{du}{dr} = A - \frac{B}{r^2} + 3Cr^2 \tag{6.74}$$

Substituting eqns. (6.73) and (6.74) in eqn. (6.66) gives

$$f_r = \frac{E}{1-\sigma^2}\left[\sigma\left(A + \frac{B}{r^2} + Cr^2\right) + A - \frac{B}{r^2} + 3Cr^2\right]$$

$$= \frac{E}{1-\sigma^2}\left[A(1+\sigma) - \frac{B}{r^2}(1-\sigma) + (\sigma+3)Cr^2\right] \tag{6.75}$$

Considering a disc with a central hole (such as a fan backplate or blade shroud plate), there will be no radial stress f_r at the inner radius r_1 or at the outer radius r_2, thus

$$0 = A(1+\sigma) - \frac{B}{r_2^2}(1-\sigma) + (\sigma+3)Cr_2^2 \tag{6.76}$$

and

$$0 = A(1+\sigma) - \frac{B}{r_1^2}(1-\sigma) + (\sigma+3)Cr_1^2 \tag{6.77}$$

which, on subtraction, give

$$(B/r_2^2 - B/r_1^2)(1-\sigma) = C(\sigma+3)(r_2^2 - r_1^2)$$

$$B\frac{r_1^2 - r_2^2}{r_1^2 r_2^2} = \frac{C(\sigma+3)(r_2^2 - r_1^2)}{(1-\sigma)}$$

$$B = -\frac{C(\sigma+3)}{(1-\sigma)} \cdot r_1^2 r_2^2 \tag{6.78}$$

Also, multiplying eqn. (6.76) by r_2^2 and eqn. (6.77) by r_1^2, and subtracting, gives

$$A(1+\sigma)(r_2^2 - r_1^2) = -C(\sigma+3)(r_2^4 - r_1^4)$$

$$A = -\frac{C(\sigma+3)}{(1+\sigma)}(r_2^2 + r_1^2) \tag{6.79}$$

Inserting these values into eqn. (6.75) gives

$$f_r = -\frac{EC}{(1-\sigma^2)}(\sigma+3)\left(r_2^2 + r_1^2 - \frac{r_1^2 r_2^2}{r^2} - r^2\right)$$

giving, on substitution for C from eqn. (6.72),

$$f_r = \frac{w \cdot \omega^2}{8g}(\sigma+3)\left(r_2^2+r_1^2-\frac{r_1^2 r_2^2}{r^2}-r^2\right) \tag{6.80}$$

This will have a maximum value when $d(f_r)/dr = 0$, that is, when

$$\frac{2r_1^2 r_2^2}{r^3} = 2r, \text{ or when } r = \sqrt{(r_1 r_2)}$$

and

$$f_{r(\text{max})} = \frac{w\omega^2}{8g}(\sigma+3)(r_2-r_1)^2 \tag{6.81}$$

Similarly, substituting the values of eqns. (6.73) and (6.74) in eqn. (6.67) gives

$$\begin{aligned}
f_h &= \frac{E}{(1-\sigma^2)}\left[A+\frac{B}{r^2}+Cr^2+\sigma\left(A-\frac{B}{r^2}+3Cr^2\right)\right] \\
&= \frac{E}{(1-\sigma^2)}\left[A(1+\sigma)+\frac{B}{r^2}(1-\sigma)+Cr^2(1+3\sigma)\right] \\
&= -\frac{CE}{(1-\sigma^2)}\left[(\sigma+3)(r_2^2+r_1^2)+\frac{r_1^2 r_2^2}{r^2}(\sigma+3)-r^2(1+3\sigma)\right] \\
&= \frac{w\omega^2}{8g}\left[(\sigma+3)\left(r_2^2+r_1^2+\frac{r_1^2 r_2^2}{r^2}\right)-r^2(1+3\sigma)\right]
\end{aligned} \tag{6.82}$$

The value of eqn. (6.82) increases as the value of r decreases, and the maximum value of f_h is when $r = r_1$.

$$\begin{aligned}
f_{h\,(\text{max})} &= \frac{w\omega^2}{8g}[(\sigma+3)(2r_2^2+r_1^2)-r_1^2(1+3\sigma)] \\
&= \frac{w\omega^2}{4g}[(\sigma+3)r_2^2+(1-\sigma)r_1^2]
\end{aligned} \tag{6.83}$$

These results should not be extrapolated for a solid disc by equating r_1 to zero. Rather, eqn. (6.73) should be solved for A and B by putting $f_r = 0$ at $r = r_1$, and $u = 0$ at $r = 0$. This results in the expected value for $f_{r(\text{max})}$ as given by eqn. (6.81) when $r_1 = 0$, but only half the value given by eqn. (6.83) when $r_1 = 0$.

6.13. STRESSING OF FAN COMPONENTS

The ideal conditions on which the previous arguments in this chapter are based rarely exist in practice. However, it is more important that a component does not fail in service than to know precisely the state of stress. The maximum recommended usable stress for most materials from which fan components are made is of the order of one-quarter to one-fifth of the ultimate breaking stress,

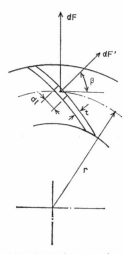

Fig. 6.21. Stress in a rotating blade.

maximum shear stress being taken as about two-thirds of the maximum permissible direct stress. Bearing this in mind, a simplified approach to the mechanical design of a fan is outlined, based on the principles of sections 6.1 to 6.12.

(a) Impeller

The backplate and shroud of a centrifugal fan may be stressed by application of eqns. (6.81) and (6.83). The effect of blade weight may be approximated by making $w =$ (disc weight + blade weight)/(disc weight) $\times$ weight per unit volume of the material used when stressing each of backplate and shroud. The blades may be considered as beams with rigid supports at backplate and shroud. From eqn. (6.53), the maximum bending moment is seen to be $Wl/12$, where W is the total distributed load on the blade, which comprises

o

centrifugal force and pressure difference across the blade. In most practical cases the centrifugal force is of greatest significance. Considering an element of blade of width b, thickness t and length dl, as shown in Fig. 6.21, the force normal to the element, dF', will be

$$dF' = dF \cos \beta = b.t.dl.\frac{w}{g}.\omega^2 r.\cos \beta$$

where w = weight per unit volume of material of the blade. Maximum bending moment,

$$M = \frac{dF'.b}{12}$$

$$= \frac{b^2 t}{12}.\frac{w}{g}.\omega^2 r.dl.\cos \beta$$

The section modulus Z will be

$$Z = \frac{t^3 dl}{12} \div \frac{t}{2} = \frac{t^2 dl}{6}$$

Thus the maximum bending stress will be

$$\frac{b^2 w.r.\omega^2 \cos \beta}{2tg} \tag{6.84}$$

If the blades are attached by rivets, these will have to withstand the total centrifugal force due to the blade. Thus shear stress on the rivets will be $W.\omega^2 r/(g.z.a)$, where W is the weight of the blade, z the number of rivets and a the cross-sectional area of each rivet. If the blades are welded to shroud and backplate, the stress in the weld will be $f = W.\omega^2 r/g.2Ll \cos 45°$, for a double fillet weld, where l is the width of the weld, since the strength of a weld is taken as being that of the weld " throat ", $l \cos 45° \times$ weld length L.

Axial flow fan impellers will also be stressed by centrifugal forces. Preferably the centroids of each element of section should lie on a radial line, when the stress at the blade root will be

$$w\omega^2 \int_{r_1}^{r_2} A(r).r.dr/(A_0 g)$$

where A_0 is the area of the blade root section and $A(r)$ is the cross sectional area of any element, often a function of the radius. Both the static pressure difference across the blade swept area and the torque will cause a combined bending moment on each blade. These should be resolved along the blade to give a bending moment at the blade root normal to a neutral axis for which the section modulus is least. The section modulus itself is probably best found by drawing an enlarged version of the aerofoil section, dividing it into a number of strip elements, and summing to give $I = \Sigma dA \times y^2$.

(b) Fan shaft

The fan shaft may be treated as a beam carrying the impellers as point loads if the shaft is long, or as a thickening of the shaft if the latter is short. The shaft must be considered in three different ways and the most rigorous of these taken as a basis of design. These are:

    (i) maximum shear stress (eqn. (6.38)),
    (ii) maximum direct stress (eqn. (6.36)),
    (iii) critical speed, which should be at least one and a half times the running speed (sections 6.8 to 6.11).

In order to carry out these processes, it will be necessary to fix the type and spacing of the bearings. Belt-driven fans will have a further load due to belt tension. Unfortunately, the true belt tension is not easy to determine, since much depends on the skill of the erector on site. It is commonly taken (with apparently little justification) as two or three times the tension required to transmit the torque to the smaller pulley.

(c) Fan casings

It is customary to strengthen, with rolled steel sections, the large expanses of metal forming the sides of centrifugal fan casings in order to avoid excessive deflections of the relatively thin sheet. The areas of sheet metal so formed may be treated as approximately rectangular plates in accordance with eqn. (6.27). The rolled steel sections are considered as beams carrying the whole load due to the pressure difference between inside and outside of the fan.

The circumferential surfaces, and also the casing of axial flow fans, may be considered as thin cylinders. The direct stress will be due to the force $p.2.r.l$ across the resisting section of area $2.t.l$ (Fig. 6.22), l being the length of casing. Thus the direct stress will be, $f = p.2.r.l/2.t.l = p.r/t$. Where an electric motor must be supported in an axial flow fan casing, this will often be suspended by tie rods from rolled steel section strengtheners fixed longitudinally to the outside of the sheet metal casing.

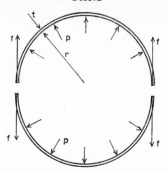

FIG. 6.22. Stress in a thin cylinder.

6.14. EXAMPLES

1. A fan impeller weighing 10 cwt has its centre of gravity acting mid-way along a shaft which is 4 ft long, 4 in. in diameter and of steel weighing 0·28 lb/in³. Show where lifting slings should be placed to limit the maximum stress due to bending during erection. Find the maximum bending stress under this condition.

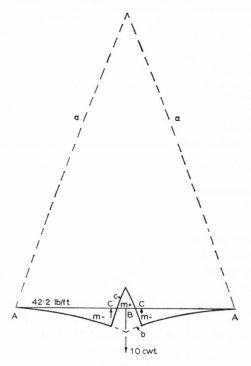

FIG. 6.23.

Referring to Fig. 6.23, the bending moment diagram for the shaft will be curve a when the slings are at points A, curve b when the slings are together at the centre of gravity of the assembly, and curve c when the slings are at points C between A and B. The minimum bending stress will occur when both of the maximum positive and maximum negative bending moments have a minimum value. Clearly, this will be when the two are equal (when $m_+ = m_-$) since a movement of the slings in either direction will result in an increase in the value of one or the other. Equating values of maximum positive and maximum negative bending moments (this will entail reversing the sign of one of them),

$$\tfrac{1}{2}wx^2 = R_C(2-x) - \tfrac{1}{2}w.2^2,$$

$$w = 0\cdot28 \times \pi/4 \times 4^2 \times 12 = 42\cdot2 \text{ lb/ft},$$

$$R_C = \tfrac{1}{2}(4 \times 42\cdot2) + \tfrac{1}{2} \times 1120 = 84\cdot4 + 560 = 644\cdot4 \text{ lb}.$$

Thus $x^2 = 2 \times 644\cdot4 \times (2-x)/42\cdot2 - 4 = 30\cdot4(2-x) - 4$

$$x^2 + 30\cdot4x - 56\cdot8 = 0$$

$$x + 15\cdot2 = \sqrt{(231 + 56\cdot8)} = 16\cdot96$$

$$x = 1\cdot76 \text{ ft}.$$

Bending moment $= \tfrac{1}{2}wx^2 = 42\cdot2 \times 1\cdot76^2/2 = 65\cdot2 \text{ lb ft}.$

Bending stress $= M/Z = \dfrac{65\cdot2 \times 12 \times 2 \times 64}{\pi \times 16^2} = 124 \text{ lb/in}^2.$

2. A fan shaft is 8 ft long, 6 in. in diameter and is made of steel weighing 480 lb/ft³. It may be taken as being simply supported at the ends. The impeller weighs 2000 lb and acts as a point load at 4 ft from one end of the shaft. It rotates at 360 rev/min and takes 500 h.p. Calculate maximum direct and shear stresses.

Weight of shaft per foot $= 480 \times \pi/(4 \times 4) = 94\cdot2 \text{ lb}.$

Total shaft weight $= 8 \times 94\cdot2 = 754 \text{ lb}.$

Reaction at either support $= \tfrac{1}{2}(754 + 2000) = 1377 \text{ lb}.$

The maximum bending moment occurs under the load and is (eqn. (6.11))
$(1377 \times 4) - (754 \times 2/2) = 4754 \text{ lb ft, or } 57,048 \text{ lb in}.$

Torque $= 33,000 \times 500/(2\pi \times 360) = 7305 \text{ lb ft, or } 87,660 \text{ lb in}.$

From eqn. (6.38), maximum shear stress $= \dfrac{16}{\pi d^3}\sqrt{(M^2 + T^2)}$

$$= \frac{16 \times 10,000}{\pi \times 216}\sqrt{(5\cdot70^2 + 8\cdot77^2)} = 235\cdot5\sqrt{109\cdot4} = 2490 \text{ lb/in}^2.$$

Maximum direct stress (eqn. (6.37)) $= \dfrac{16}{\pi d^3}[M + \sqrt{(M^2 + T^2)}]$

$$= \frac{16 \times 10,000}{\pi \times 216}[5\cdot70 + \sqrt{(5\cdot70^2 + 8\cdot77^2)}] = 3810 \text{ lb/in}^2.$$

3. A steel shaft, 3 in. in diameter, is 5 ft long, and may be considered as being simply supported at the ends. It carries two impellers, each weighing 400 lb, the centres of gravity of which are 1 ft and 3 ft respectively from one end of the shaft. Calculate the first critical speed of the system.

Weight of shaft $= 5 \times \dfrac{\pi}{4} \times \dfrac{3^2}{144} \times 480 = 117 \cdot 9$ lb.

Second moment of area of shaft $= \pi d^4/64$ (eqn. (6.23))
$= \pi \times 81/64 = 3 \cdot 97$ in^4.

Deflection due to weight of shaft $= \dfrac{5wl^4}{384EI}$ (eqn. (6.45))

$$y_{max} = \frac{5 \times 117 \cdot 9 \times 60 \times 60 \times 60}{384 \times 30 \times 10^6 \times 3 \cdot 97} = 2 \cdot 79 \times 10^{-3} \text{ in.}$$

This occurs at the mid-point of the shaft.

Deflection under the first impeller $= \dfrac{Wa^2b^2}{3EIl}$ (eqn.(6.43))

$$y_{01} = \frac{400 \times 12^2 \times 48^2}{3 \times 30 \times 10^6 \times 3 \cdot 97 \times 60} = 6 \cdot 19 \times 10^{-3} \text{ in.}$$

Deflection under the second impeller

$$y_{02} = \frac{400 \times 36^2 \times 24^2}{3 \times 30 \times 10^6 \times 3 \cdot 97 \times 60} = 13 \cdot 93 \times 10^{-3} \text{ in.}$$

From eqns. (6.61) and (6.60), $1/f_0^2 = 1/f_{01}^2 + 1/f_{02}^2 + 1/f_{0w}^2$ and

$f_{01} = \dfrac{1}{2\pi} \sqrt{\dfrac{g}{y_{01}}}$ and $f_{0w} = \dfrac{1}{2\pi} \sqrt{\left(\dfrac{1 \cdot 27g}{y_{max}}\right)}$, from which

$$y_0 = y_{01} + y_{02} + y_{max}/1 \cdot 27$$
$$= (6 \cdot 19 + 13 \cdot 93 + 2 \cdot 79/1 \cdot 27) \times 10^{-3} = 22 \cdot 32 \times 10^{-3} \text{ in.}$$

$$f_0 = \frac{1}{2\pi} \sqrt{\frac{g}{y_0}} = \frac{1}{2\pi} \sqrt{\left(\frac{32 \cdot 2 \times 12 \times 10^3}{22 \cdot 32}\right)} = 20 \cdot 91 \text{ c/s.}$$

This is the fundamental frequency of transverse vibration of the system and the first critical speed will be $20 \cdot 91 \times 60 = 1255$ rev/min.

4. A fan shaft, 8 ft long, is 4 in. in diameter for a length of 1 ft at each end, and 6 in. in diameter for the middle 6 ft. It carries an impeller weighing 2000 lb at its mid-point. Find, by a graphical method, the diagram of deflection.

Weight per inch length of 6 in. diameter shaft (of steel)

$$= \frac{\pi \times 6^2 \times 480}{4 \times 1728} = 7 \cdot 85 \text{ lb/in.}$$

Weight per inch length of 4 in. diameter shaft

$$= \frac{\pi \times 4^2 \times 480}{4 \times 1728} = 3 \cdot 49 \text{ lb/in.}$$

Total shaft weight $= 24 \times 3 \cdot 49 + 72 \times 7 \cdot 85 = 648 \cdot 7$ lb.
Reaction at either support $= \frac{1}{2}(648 \cdot 7 + 2000) = 1324 \cdot 3$ lb.

Bending moment along the shaft from either end:

At 1 ft, $M_1 = 1324 \times 12 - 3\cdot49 \times 12 \times 6 = 15,892 - 251 = 15,641$ lb in.

At 2 ft, $M_2 = 1324 \times 24 - 3\cdot49 \times 12 \times 18 - 7\cdot85 \times 12 \times 6$
$= 31,784 - 753 - 564 = 30,467$ lb in.

At 3 ft, $M_3 = 1324 \times 36 - 3\cdot49 \times 12 \times 30 - 7\cdot85 \times 24 \times 12$
$= 47,676 - 1257 - 2260 = 44,159$ lb in.

At 4 ft, $M_4 = 1324 \times 48 - 3\cdot49 \times 12 \times 42 - 7\cdot85 \times 36 \times 18$
$= 63,568 - 1756 - 5085 = 56,727$ lb in.

Second moment of area:

For 4 in. diameter shaft, $I_1 = \pi \times 4^4/64 = 12\cdot6$ in^4.

For 6 in. diameter shaft, $I_2 = \pi \times 6^4/64 = 63\cdot6$ in^4.

At 1 ft, $M_1/I_1 = 15,641/12\cdot6 = 1241$ lb in^{-3}.
$M_1/I_2 = 15,641/63\cdot6 = 246$ lb in^{-3}.

At 2 ft, $M_2/I_2 = 30,467/63\cdot6 = 480$ lb in^{-3}.

At 3 ft, $M_3/I_2 = 44,159/63\cdot6 = 695$ lb in^{-3}.

At 4 ft, $M_4/I_2 = 56,727/63\cdot6 = 892$ lb in^{-3}.

The other half of the shaft is symmetrical.

The " areas " of the M/I diagram (Fig. 6.24) are taken rather crudely at 12 in. intervals in order to clarify the example.

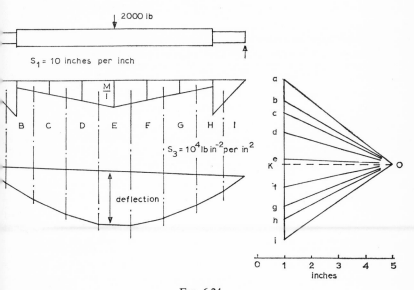

FIG. 6.24.

They will have the units of in. $\times$ lb in^{-3}, or lb in^{-2}.

Area 1 $= \frac{1}{2} \times 1241 \times 12 = 7446$ lb in^{-2}.

Area 2 $= \frac{1}{2} \times (246 + 480) \times 12 = 4356$ lb in^{-2}.

Area 3 $= \frac{1}{2} \times (480 + 695) \times 12 = 7050$ lb in^{-2}.

Area 4 $= \frac{1}{2} \times (695 + 892) \times 12 = 9522$ lb in^{-2}.

Areas 5, 6, 7 and 8 have the same values as areas 4, 3, 2 and 1 respectively.

These " areas " are plotted as a diameter of forces would be plotted for construction of a bending moment diagram and result in the vertical line on the right-hand side of Fig. 6.24. A pole K is selected and rays drawn to the appropriate points on the vertical line. Lines parallel to these rays are drawn in the corresponding spaces between centroids of adjacent areas of the M/I diagram (the line in space A being parallel to ray oa, and so on), the resulting diagram showing the deflection of the beam to a scale of

$$\frac{1 \times S_1 S_2 OK}{E} = \frac{1 \times 10 \times 10^4 \times 4}{30 \times 10^6} = 1 \cdot 33 \times 10^{-2} \text{ in.}$$

of deflection per inch of diagram measured vertically.

5. A fan backplate of steel has an outer diameter of 2 ft and an inner diameter of 6 in., and rotates at 1800 rev/min. It carries blades of total weight equal to the backplate weight. Assuming the effect of the blades to be equivalent to an increase in weight per unit volume of the backplate of 50%, calculate the maximum radial and hoop stresses. Take $\sigma = 0.29$.

Weight per unit volume of backplate with allowance for blade weight

$$= 1 \cdot 5 \times 480 = 600 \text{ lb/ft}^3.$$

Angular velocity of rotation, $\omega = 2\pi \times 1800/60 = 188 \cdot 5$ rad/s.

Maximum radial stress $= \dfrac{w\omega^2}{8g} (\sigma + 3)(r_2 - r_1)^2$ (eqn. (6.81))

$$= \frac{600 \times 188 \cdot 5^2}{8 \times 32 \cdot 2} \times 3 \cdot 29 \times (1 - 0 \cdot 25)^2 = 153{,}500 \text{ lb/ft}^2$$

$$= 1065 \text{ lb/in}^2.$$

Maximum hoop stress $= \dfrac{w\omega^2}{4g} [(\sigma + 3)r_2^2 + (1 - \sigma)r_1^2]$

$$= \frac{600 \times 188 \cdot 5^2}{4 \times 32 \cdot 2} \times [3 \cdot 29 \times 1^2 + 0 \cdot 71 \times 0 \cdot 25^2] = 552{,}000 \text{ lb/ft}^2$$

$$= 3830 \text{ lb/in}^2.$$

APPENDIX I

PRELIMINARY design of a centrifugal fan for a duty of 8000 ft³/min at a fan static pressure of 2·0 in. of water under standard air conditions.

From Table 5.1, it is seen that a 12-bladed backward curved fan, for which the outlet blade angle β_2 is 40°, would have a total efficiency of at least 75%. It is proposed to use this as a basis of the design.

Taking outlet velocity $v_3 = 0·3u_2$ and using the conclusions of section 5.3

$$\frac{p_s + \frac{1}{2}\rho(0·3u_2)^2}{\eta} = \rho u_2\left(u_2 - \frac{\pi \sin \beta_2}{z} - v_m \cot \beta_2\right)$$

$$\frac{2·0 \times 32·2 \times 5·2 + \rho \times 0·045u_2^2}{0·75} = \rho u_2^2\left(1 - \frac{\pi \times 0·643}{12} - 0·2 \times 1·192\right)$$

$$u_2 = \sqrt{\left(\frac{2·0 \times 32·2 \times 5·2}{0·75 \times 0·075 \times 0·534}\right)} = 112 \text{ ft/s.}$$

From Table 5.1, a reasonable value for ϕ would be 0·18:

$$0·18 = \frac{4 \times Q}{\pi d_2^2 u_2}$$

$$d_2 = \sqrt{\left(\frac{4 \times 8000}{\pi \times 0·18 \times 112 \times 60}\right)} = 2·9 \text{ ft.}$$

Speed of rotation

$$n = \frac{112 \times 60}{\pi \times 2·9} = 740 \text{ rev/min.}$$

Now $v_{m1} = v_{m2} = 0·2u_2$, say $= 22·4$ ft/s,

$$v_0 = 2v_{m1} = 44·8 \text{ ft/s.}$$

201

Impeller inlet diameter

$$d_1 = \sqrt{\left(\frac{4Q}{\pi d_1^2 v_0}\right)} = \sqrt{\left(\frac{4 \times 8000}{\pi \times 8{\cdot}42 \times 44{\cdot}8 \times 60}\right)}$$

$$= 2{\cdot}12 \text{ ft.}$$

Impeller widths b_1 and b_2 may now be found:

$$b_1 = \frac{Q}{\pi d_1 v_{m1}} = \frac{8000}{\pi \times 2{\cdot}12 \times 22{\cdot}4 \times 60} = 0{\cdot}9 \text{ ft,}$$

$$b_2 = \frac{d_1}{d_2} b_1 \text{ (since } v_{m2} = v_{m1}) = \frac{2{\cdot}12}{2{\cdot}9} \times 0{\cdot}9 = 0{\cdot}73 \text{ ft.}$$

Inlet blade angle,

$$\beta_1 = \tan^{-1} \frac{v_{m1}}{u_1} = \tan^{-1}\left(\frac{22{\cdot}4 \times 2{\cdot}9}{2{\cdot}12 \times 112}\right) = 15{\cdot}3°.$$

Casing outlet velocity, $v_3 = 0{\cdot}3u_2 = 0{\cdot}3 \times 112 = 33{\cdot}6$ ft/s.

Casing width $= 2{\cdot}5b_2$ (say) $= 2{\cdot}5 \times 0{\cdot}73 = 1{\cdot}83$ ft.

Casing outlet area $= \dfrac{8000}{33{\cdot}6 \times 60} = 3{\cdot}97$ ft^2.

Casing outlet height $= 3{\cdot}97/1{\cdot}83 = 2{\cdot}17$ ft.

Making casing inlet area equal to the casing outlet area, the casing inlet diameter $= \sqrt{(4 \times 3{\cdot}97/\pi)} = 2{\cdot}25$ ft.

Preliminary stress calculations are now carried out. It is necessary to estimate the weight of the impeller, which will be made from $\frac{1}{4}$ in. mild steel plate of 490 lb/ft^3.

Backplate weight $= \pi/4 \times 2{\cdot}9^2 \times 490/48 = 67{\cdot}5$ lb.

Shroud weight $= \pi/4 \times (2{\cdot}9^2 - 2{\cdot}12^2) \times 490/48 = 31$ lb.

Considering each blade as having a length of 0·6 ft and a width of $\frac{1}{2}(0{\cdot}9 + 0{\cdot}73) = 0{\cdot}82$ ft.

Total blade weight $= 12 \times 0{\cdot}6 \times 0{\cdot}82 \times 490/48 = 60$ lb.

The total length of fillet weld will probably be $2 \times 2 \times 12 \times 0{\cdot}6 = 29$ ft.

Weight of fillet weld

$$= 29 \times \frac{1}{4\sqrt{2}} \times \frac{490}{144} = 18 \text{ lb.}$$

(This has been overestimated a little.) The hub weight is estimated as 20 lb. Thus the total impeller weight is 197 lb, say 200 lb.

Considering the stress in the backplate (which will be greater than for the shroud), and allowing half the blade weight to be effective, the corrected weight per unit volume will be $490 \times (67 \cdot 5 + 30)$ $67 \cdot 5$ $= 708$ lb/ft^3 and σ will be about $0 \cdot 29$.

Maximum radial stress, $f_r = w\omega^2[(\sigma+3)r_2^2]/8g$ from eqn. (6.81) and considering r_1 to be very small,

$$f_r = \frac{708}{8 \times 32.2}\left(\frac{112}{1 \cdot 45}\right)^2 \times 3 \cdot 29 \times (1 \cdot 45)^2 = 113{,}200 \text{ lbf/ft}^2$$
$$= 925 \text{ lbf/in}^2.$$

Maximum hoop stress,

$$f_h = \frac{w\omega^2}{4g}[(\sigma+3)r_2^2] \quad r_1 \to 0$$

$$= \frac{708}{4 \times 32 \cdot 2} \times \left(\frac{112}{1 \cdot 45}\right)^2 \times 3 \cdot 29 \times (1 \cdot 45)^2 = 226{,}400 \text{ lbf/ft}^2$$
$$= 1850 \text{ lbf/in}^2$$

The stresses are seen to be quite low. In the blade shroud, the stresses will clearly be of the same order since r_1 is greater, although w becomes $490 \times (31 + 30)/31 = 965$ lb/ft^3.

To design the shaft it is necessary to make an estimate of the overhang of the impeller from the nearest bearing. Taking this to be 8 in., the bending moment $= 200 \times 8 = 1600$ lbf in.

Power required to drive the fan

$$= \frac{8000 \times 2 \cdot 253 \times 5 \cdot 2}{33{,}000 \times 0 \cdot 75} = 3 \cdot 78 \text{ h.p.}$$

Torque

$$= \frac{33{,}000 \times 3 \cdot 78 \times 12}{2 \times \pi \times 740} = 323 \text{ lbf in.}$$

Based on a maximum direct stress (eqn. (6.37)) of 18,000 lbf/in^2,

$$d^3 = \frac{16}{\pi \times 18{,}000}[1600 + \sqrt{(1600^2 + 323^2)}] = 0 \cdot 915$$

$$d = 0 \cdot 97 \text{ in.}$$

Based on maximum shear stress (eqn. (6.38)) of 12,000 lbf/in^2,

$$d^3 = \frac{16}{\pi \times 12,000}[\sqrt{(1600^2 + 323^2)}] = 0\cdot693$$

$$d = 0\cdot89 \text{ in.}$$

It is further necessary to find the minimum diameter to give an acceptable critical speed of the system. Taking this speed to be at least $1\cdot5 \times 740 = 1120$ rev/min, and referring to section 6.10, the deflection of the impeller resulting in a natural frequency of transverse vibration of $1120/60$ c/s is

$$y = \left(\frac{3\cdot13}{f_0}\right)^2 = \left(\frac{3\cdot13 \times 60}{1120}\right)^2 = 0\cdot028 \text{ in.}$$

Considering the impeller as a point load at the free end of the shaft (as a cantilever), 8 in. from a rigid bearing, and applying eqn. (6.41), deflection

$$y = \int \left[\int \frac{M}{EI} dx + C \right] dx + D$$

Taking x from the free end,

$$M_x = Wx$$

$$i_x = \frac{1}{EI}\left[\frac{Wx^2}{2} + C \right]$$

$$y_x = \frac{1}{EI}\left[\frac{Wx^3}{6} + Cx + D \right]$$

Since $i_x = 0$ at $x = l$, $C = -wl^2/2$; and since $y_x = 0$ at $x = l$, $D = Wl^2/2 - Wl^3/6 = Wl^3/3$.

Thus the deflection at the free end where $x = 0$,

$$y_0 = \frac{Wl^3}{3EI}$$

and

$$0.028 = \frac{200 \times 8^3 \times 64}{3 \times 30 \times 10^6 \times \pi d^4}$$

$$d^4 = \frac{7.91 \times 10^6}{6.55 \times 10^6} = 1.21$$

$$d = 1.05 \text{ in}$$

From these three considerations it is clear that the shaft diameter should not be less than about 1 in. A fan such as this might form part of a series of fans and may be called upon to run at rather higher speeds than 740 rev/min. Allowing for this and for the fact that the conditions may be a little more complex than have been assumed, the shaft diameter would be increased, thus relieving the stress due to bending and torsion. A fan of this size would commonly have a shaft diameter of $2\frac{1}{2}$ in.

Checking the blade bending stress from eqn. (6.84)

$$f = \frac{b^2 w r \omega^2}{2tg} \cos \beta$$

At radius $r_2 = 1.45$ ft, $\beta = 40°$, $b_2 = 0.73$ ft,

$$f = \frac{0.73^2 \times 490 \times 1.45 \times 6000 \times 48}{2 \times 32.2 \times 144} \times 0.766 = 9000 \text{ lbf/in}^2.$$

At radius $r_1 = 1.06$ ft, $\beta = 15.3°$, $b_1 = 0.9$ ft,

$$f = \frac{0.9^2 \times 490 \times 1.06 \times 6000 \times 48}{2 \times 32.2 \times 144} \times 0.97$$

$$= 13,000 \text{ lbf/in}^2.$$

These stresses are acceptable, if high. Checking on the weld strength, weld length per blade = $4 \times 0.6 = 2.4$ ft.
Weld throat area = $2.4 \times 12 \times \frac{1}{4} \times (1/\sqrt{2}) = 5.1$ in^2.
Centrifugal force per 5 lb blade rotating at average radius 1.26 ft and rotating at 740 rev/min

$$= \frac{w}{g} \omega^2 r = \frac{5 \times 6000 \times 1.26}{32.2} = 1175 \text{ lb.}$$

Average weld stress = $1175/5.1 = 230$ lbf/in^2.

Finally, a check on casing thickness. This would be contained within an area 5 ft by 5 ft

$$t^2 = \frac{pa^2b^2}{2f(a^2+b^2)} = \frac{2\cdot253 \times 5\cdot2 \times 25 \times 25}{2 \times 18{,}000 \times 50}$$

$$= 4\cdot07 \times 10^{-3}$$

$$t = 6\cdot4 \times 10^{-2} \text{ in.}$$

Considerations of rigidity, handling and corrosion would invariably result in much greater thickness than this.

APPENDIX II

PRELIMINARY design of a downstream guide vane axial flow fan for a duty of 5000 ft^3/min at a fan total pressure of 0·5 in. of water under standard air conditions.

From Table 5.2, it is seen that, for an efficiency of 80% a downstream guide vane axial flow fan could have values of, for example, $\psi = 0·35$ and $\phi = 0·28$ with a hub ratio ν of 0·5.

The design total pressure will be 0·5/0·8 = 0·625 in. water.

$$\tfrac{1}{2}\rho u^2 = p_t/\psi = 0·625/0·35 = 1·785 \text{ in. water.}$$

$$u = \sqrt{\left(\frac{1·785 \times 5·2 \times 32·2 \times 2}{0·075}\right)} = 89·2 \text{ ft/s.}$$

Axial flow fans are usually directly driven by electric motors which, on a.c. supply frequency of 50 c/s, will have speeds of 2900, 1450 or 975 rev/min for low powers. Basing the design on a motor speed of 975 rev/min, or 16·3 rev/s, the diameter will be

$$d = u/\pi n = 89·2/(\pi \times 16·3) = 1·75 \text{ ft.}$$

If this diameter were selected, the fan discharge area would be $\pi/4 \times (1·75)^2 = 2·4$ ft^2. The fan velocity pressure would be $[5000/(2·4 \times 4000)]^2 = 0·27$ in. water and the resulting fan static pressure would be $0·5 - 0·27 = 0·23$ in. of water. This is a rather high ratio of fan static to fan total pressure. A fan diameter of 2 ft gives a discharge area of 3·14 ft^2 and a velocity pressure of $[5000/(3·14 \times 4000)]^2 = 0·16$ in. water resulting in a fan static pressure of 0·34 in. of water, a rather more reasonable value. Proceeding with the design for a fan having a tip diameter of 2 ft and a hub diameter of 1 ft:

$$v_m = \frac{5000 \times 4}{60 \times \pi(2^2 - 1^2)} = 35·4 \text{ ft/s.}$$

207

Starting the design at the hub section:

$$u = \pi \times 1 \times 16\cdot3 = 51\cdot2 \text{ ft/s,}$$

$$pt = \rho u v_u; \quad v_u = pt/\rho u$$

$$= \frac{0\cdot625 \times 5\cdot2 \times 32\cdot2}{0\cdot075 \times 51\cdot2} = 27\cdot2 \text{ ft/s.}$$

$$\beta - \alpha = \tan^{-1}(v_m/u - v_u/2) = \tan^{-1}(35\cdot4/51\cdot2 - 13\cdot6) = 43°13'.$$

$$w_\infty = \sqrt{[v_m^2 + (u - v_u/2)^2]} = \sqrt{(35\cdot4^2 + 37\cdot6^2)} = 51\cdot6 \text{ ft/s.}$$

$$v_u/w_\infty = 27\cdot2/51\cdot6 = 0\cdot528.$$

From single aerofoil theory it would be just possible to design this section. Using a Göttingen 436 aerofoil, some data for which are given in Fig. 1.13, and a value of $c/s = 1$, the value of C_L from eqn. (5.40) is $2 \times 0\cdot528/1 = 1\cdot056$ corresponding to an angle of attack of about 6°. Thus the root blade angle β will be about $43° + 6° = 49°$ (bearing in mind that the accuracy of setting the blade angle will generally be $\pm 0\cdot5°$).

Checking this against cascade data:

$$\alpha_1 = \tan^{-1} u/v_m = \tan^{-1} 51\cdot2/35\cdot4 = 55\cdot3°$$

$$\tan \alpha_2 = \tan \alpha_1 - v_u/v_m = 1\cdot45 - 27\cdot2/35\cdot4 = 0\cdot66$$

$$\alpha_2 = 33\cdot4°.$$

From Fig. 5.17 it will be seen that these values correspond to a value of c/s of rather less than unity.

The fluid deflection in the guide vane will be $\tan^{-1} v_u/v_m = 37\cdot5°$. Using aerofoil section guide vanes, the necessary value of C_L may be found from eqn. (5.50). Neglecting drag,

$$C_L = (2s/c)(\tan \alpha_1 - \tan \alpha_2) \cos \alpha_m$$

Taking $\alpha_1 = 37\cdot5°$, $\alpha_2 = 0°$ and $\alpha_m = 18°$,

$$C_L = (2s/c) \times 0\cdot7673 \times 0\cdot9511 = 1\cdot46s/c$$

This is higher than can be achieved by aerofoils with solidity of unity from single aerofoil theory. Altering the angle of the aerofoil to give $\alpha_1 = 45°$, $\alpha_2 = 9\cdot5°$ and $\alpha_m = 27\cdot5°$ gives

$$C_L = (2s/c) \times (1 - 0\cdot1673) \times 0\cdot887 = 1\cdot47s/c$$

which is much the same. Reference to Fig. 5.17 for cascade data shows that this deflection can be achieved with a value of c/s of the order of 1·2–1·3.

The tip section of the blade may now be designed.

$$u = 2 \times 51\cdot2 = 102\cdot4 \text{ ft/s}$$

$$v_u = 27\cdot2/2 = 13\cdot6 \text{ ft/s} \quad (\text{taking } v_u \cdot r = \text{constant})$$

$$u - v_u/2 = 102\cdot4 - 6\cdot8 = 95\cdot6$$

$$\beta - \alpha = \tan^{-1} 35\cdot4/95\cdot6 = 20°18'$$

$$w_\infty = \sqrt{(35\cdot4^2 + 95\cdot6^2)} = 101\cdot5 \text{ ft/s}$$

$$v_u/w_\infty = 13\cdot6/101\cdot5 = 0\cdot134$$

$$C_L c/s = 2 \times 0\cdot134 = 0\cdot268.$$

In selecting the blade tip solidity, consideration is given to the saving of material in manufacture and a chord length selected with this in view, bearing in mind preservation of accuracy of manufacture of the aerofoil profile. For a 9-bladed fan with a hub solidity of unity the blade chord at the hub will be $12 \times \pi/9 = 4\cdot2$ in. A tip chord of 3 in. would not be unreasonable and thus, at the tip

$$c/s = \frac{9 \times 3}{24 \times \pi} = 0\cdot36$$

from which C_L would be $0\cdot268/0\cdot36 = 0\cdot745$. From Fig. 1.13, α is seen to be about 2·4°. Thus $\beta = 20\cdot3 + 2\cdot4 = 22\cdot7°$.

The fluid deflection in the guide vane will be

$$\tan^{-1} v_u/v_m = \tan^{-1} 13\cdot6/35\cdot4 = 21°$$

Using eqn. (5.50) and taking α_m as being about 10·5°, and neglecting C_D

$$C_L c/s = 2(\tan \alpha_1 - \tan \alpha_2) \cos \alpha_m = 2 \times 0\cdot384 \times 0\cdot983$$

$$= 0\cdot754$$

The number of guide vanes might reasonably be one more, or one less, than the number of fan blades and should be spaced by at least half of the chord from the blades to avoid excessive noise. Selecting a value of C_L which is about the same as for the impeller blades will enable the guide vane chords to be determined.

P

The foregoing procedure may be repeated for a number of sections between hub and tip as desired.

Preliminary stress calculations should now be performed. Consider first the stresses in the fan blades due to centrifugal force. Assuming the impeller to be cast in one piece in aluminium ($w = 160$ lb/ft^3) and that the blades are designed such that the profile centroids lie on a radial line, there will be a direct stress on the root section. The area of section at any radius r is $A =$ constant $\times c^2$, where chord c is a function of the radius. If the blades have a straight taper,

$$c = c_1 - \frac{r - r_1}{r_2 - r_1}(c_1 - c_2)$$

If $c_1 = 0.35$ ft at $r_1 = 0.5$ ft and $c_2 = 0.25$ ft at $r_2 = 1.0$ ft,

$$c = 0.35 - \left(\frac{r - 0.5}{1 - 0.5}\right) \times 0.1 = 0.35 - 0.2(r - 0.5)$$

$$= \tfrac{1}{10}(3.5 - 2r + 1) = \tfrac{1}{10}(4.5 - 2r)$$

Stress

$$= \frac{w\omega^2}{gA_1}\int_{r_1}^{r_2} A\,dr.r = \frac{160 \times (16.3 \times 2\pi)^2}{144 \times 32.2 \times 0.35^2}$$

$$\times \int_{r_1 = 0.5}^{r_2 = 1}\left(\frac{20.25r - 18r^2 + 4r^3}{100}\right)\,dr \text{ lbf/in}^2$$

$$= 29.5[10.13r^2 - 6r^3 + r^4]_{0.5}^{1}$$

$$= 29.5 \times 3.28 = 97 \text{ lbf/in}^2.$$

The combined effect of lift and torque forces on the blade, if drag force is ignored, is the lift force, $\int p_s'.2\pi r dr/\cos{(\beta - \alpha)}$ (section 5.9). The bending moment resulting from this will be (for z blades)

$$M_z = \int_{r_1}^{r_2}\frac{p_s'2\pi r^2 dr}{\cos{(\beta - \alpha)}}$$

If the blade is not severely stressed, p_s' may be considered as being the fan total pressure initially, and $\cos(\beta - \alpha)$ may be taken as being at the blade root (having the least value at this point). This will overestimate slightly the bending moment

$$M_z = \frac{2\pi p_t}{3\cos(\beta - \alpha)}(r_2^3 - r_1^3)_{0.5}^1 = \frac{2\pi p_t}{3\cos(\beta - \alpha)} \times 0.875$$

$$= \frac{1.83 p_t}{\cos(\beta - \alpha)}$$

The bending moment per blade will be, for 9 blades, $0.203 p_t/\cos(\beta - \alpha)$ and for $\cos(\beta - \alpha) = 0.73$ and

$$p_t = 0.625 \times 5.2 \text{ lbf/ft}^2$$

$$M = \frac{0.203 \times 0.625 \times 5.2}{0.73} = 0.89 \text{ lbf ft}$$

$$= 10.7 \text{ lbf in.}$$

Referring to Fig. A, the approximate section modulus for a Göttingen 436 profile may be found in terms of the chord length c.

Section	Area A $10^{-3} c^2$	h $10^{-2} c$	y $10^{-2} c$	Ay $10^{-5} c^3$	Ah^2 $10^{-7} c^4$
1	6.4	5.5	10	64	193
2	9.8	3.5	8	78.4	118
3	13.4	1.5	6	80.4	30
4	15.8	0.5	4	63.1	4
5	26.7	3.0	1.5	40	239
	$\Sigma A = 72.1$			$\Sigma Ay = 325.9$	$I = 584$

Position of centroid from undersurface

$$= \Sigma Ay/\Sigma A = \frac{325.9 \times 10^{-5} c^3}{72.1 \times 10^{-3} c^2}$$

$$= 4.5 \times 10^{-2} c$$

Distance from centroid to farthest material fibre

$$= (11 - 4.5) \times 10^{-2} c = 6.5 \times 10^{-2} c$$

212 FANS

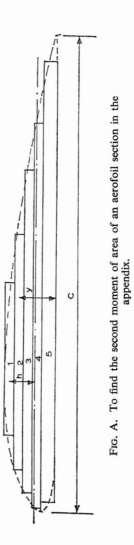

Fig. A. To find the second moment of area of an aerofoil section in the appendix.

Smallest section modulus

$$= (5{\cdot}84 \times 10^{-5}c^4)/(6{\cdot}5 \times 10^{-2}c) = 0{\cdot}9 \times 10^{-3}c^3$$

In this design

$$c = 4{\cdot}2 \text{ in., thus } z = 0{\cdot}9 \times 4{\cdot}2^3 \times 10^{-3} = 66{\cdot}5 \times 10^{-3} \text{in}^3.$$

Bending stress $= 10{\cdot}7 \times 10^3/66{\cdot}5 = 161$ lbf/in^2 making the total direct stress $161 + 97 = 258$ lbf/in^2.

If the shaft size and overhang for the motor drive is known, the critical speed may be calculated as for the centrifugal fan design.

The minimum casing thickness for a stress of 18,000 lbf/in^2 will be $t = pr/f = 0{\cdot}625 \times 5{\cdot}2 \times 12/18{,}000 = 1{\cdot}8 \times 10^{-3}$ in. The casing would, of course, be made much thicker than this.

LIST OF REFERENCES

1. British Standard 848: Part 1: 1963. British Standards Institution, London.
2. OWER, E., *The Measurement of Air Flow*, 3rd edition, Chapman & Hall, London, 1949.
3. WINTERNITZ, F. A. L. and FISCHL, C. F., *Water Power*, **9**, 225 (1957).
4. NIKURADSE, J., *V.D.I. Forschungshefte*, 361 (1933).
5. COLEBROOK, C. F., *J.I.C.E.* **2**, 133 (1938).
6. MOODY, L. F., *Trans. A.S.M.E.* **66**, 671 (1944).
7. *A Guide to Current Practice*, 1959. Institution of Heating and Ventilating Engineers, London.
8. A.S.H.R.A.E. Guide and Data Book, *Fundamentals and Equipment*, A.S.H.R.A.E., New York, 1964.
9. GIBSON, A. H., *Hydraulics and its Applications*, 4th edition, Constable, London, 1930.
10. HAY, D., *Trans. Inst. Min. E.* **67** (part 3), 268 (1924).
11. TAYLOR, S. N. (communicated). See also *The Services Textbook of Radio*, Vol. 1, p. 131, H.M.S.O., London, 1956.
12. DALY, B. B., *G.E.C. Journal*, **13** (1944).
13. ECK, B., *Ventilatoren*, Springer-Verlag, Berlin, 1962.
14. HAGEN, H. F. (In discussion of paper by M. S. KICE.) *J.I. Mech. E.* **154**, 115 (1946).
15. CURLE, N., *Proc. Roy. Soc.* A **231**, 505 (1955).
16. CHADDOCK, J. B. et al. *Refrigerating Engineering*, **67**, 37 (1959).
17. BERANEK, L. L., *Noise Reduction*, McGraw-Hill, New York, 1960.
18. STODOLA, A., *Dampf- und Gasturbinen*, 6 Aufl, Springer, Berlin, 1924.
19. KEARTON, W. J., *J.I. Mech. E.* **124**, 481 (1933).
20. HOWELL, A. R., *Proc. I. Mech. E.* **153**, 441 (1945).
21. British Standard 4: Part 1: 1962. British Standards Institution, London.
22. DUNKERLEY, S., *Phil. Trans. Roy. Soc.* A **185**, 279 (1894).

FURTHER READING

AEROFOILS

ABBOT, I. H. and VON DOENHOFF, A. B., *Theory of Wing Sections*, McGraw-Hill, New York, 1949.

CARTER, A. D. S., *The Low Speed Performance of Related Aerofoils in Cascade*, Aero. Res. Council C.P. 29 (1950).

CARTER, A. D. S., Blade profiles for axial flow fans, pumps, compressors, *Proc. I. Mech. E.* **175**, 15, 775–806 (1961).

GLAUERT, H., *The Elements of Aerofoil Theory*, 2nd Edition, Cambridge University Press, 1948.

LIEBLEIN, S., Incidence and deviation angle corrections for compressor cascades, *Trans. A.S.M.E.* Ser. D, **82**, 575–87 (1960).

PANKHURST, R. C., *N.P.L. Aerofoil Catalogue and Bibliography*, Aero. Res. Council, R. & M. 3311 (1961).

RIEGELS, F. W., *Aerofoil Sections*, Butterworth, London, 1961.

SCHLICTING, H., Application of boundary layer theory in turbomachinery, *Trans. A.S.M.E.* Ser. D, **81**, 543–51 (1959).

TRAUPEL, W., Calculation of potential flow through blade grids, *Sulzer Review* No. 1, pp. 25–42 (1945).

AXIAL FLOW FANS

BARNA, P. S., Estimation of scale effects in axial flow fans, *Aircraft Eng.* **33**, 393, 314–20 (1961).

BRAMLEY, R. (Mrs.) and McCOY, E., *Bibliography on Axial Flow Compressors*, U.K.A.E.A. Deg. 80(CA) (1960).

FUJIE, F., A study of the flow through the rotor of an axial compressor, *Bull. Japan. S.M.E.* **5**, 18, 292–301 (1962).

HAWTHORNE, W. R. and HORLOCK, J. H., Actuator disc theory of the incompressible flow in axial compressors, *Proc. I. Mech. E.* **176**, 130, 789–814 (1962).

HORLOCK, J. H., *Instrumentation Used in the Measurement of the Three-dimensional Flow of an Axial Compressor Stage*, Aero. Res. Council C.P. 321.

HOWELL, A. R., *The Present Basis of Axial Flow Compressor Design*, Aero. Res. Council, R. & M. 2095 (1942).

HOWELL, A. R., Fluid dynamics of axial compressors, *Proc. I. Mech. E.* W.E. 12, **153**, 441–52 (1945).

HOWELL, A. R., Design of axial compressors, *Proc. I. Mech. E.* **153**, 452–62 (1945).

HUTTON, S. P., Three-dimensional motion in axial flow impellers, *Proc. I. Mech. E.* **170**, 25, 863–73 and 888–908 (1956).

LAKSHMINARAYANA, B. and HORLOCK, J. H., *Tip Clearance Flow and Losses for an Isolated Compressor Blade*, Aero. Res. Council, R. & M. 3316 (1962).

LYNAM, F. C. and HAWES, S. P., Contra-rotating axial flow fans, *Engineer*, July 1946.

MAIR, W. A., *The Design of Fans and Guide Vanes for High Speed Wind Tunnels*, Aero. Res. Council, R. & M. 2435 (1951).

MARCINOWSKI, H., Optimisation problems of axial flow fans (in German), *Heiz.-Luft.-Haustech.* **8,** 11, 273–85, 295–6 (1957).

MARPLES, J., The design of an axial flow fan and a discussion on experimental results, *Journal I.H.V.E.* **24,** 361–74 (1956).

MIKHAIL, S., Three-dimensional flow in axial pumps and fans, *Proc. I. Mech. E.* **172,** 35, 973–96 (1958).

MYLES, D. J. and WATSON, J. T. R., *The Design of Axial Flow Fans by Computer* (part 1), N.E.L. Rep. 145 (1964).

POOLE, R., *The Theory and Design of Propeller Type Fans*, Inst. C.E. Selected papers 1178 (1935).

RAKOCZY, T., The optimum arrangement of the impeller and guide vanes in axial flow fans (in German), *Heiz.-Luft.-Haustech.* **13,** 6, 165–70; 9, 295–300; 10, 371–2 (1962).

SEIDEL, B. S., Assymetric inlet flow in axial turbomachinery, *Trans. A.S.M.E.* Ser. A, **86,** 1, 18–28 (1964).

SEROVY, G. K. and LYSEN, J. C., Prediction of axial flow turbomachinery performance by blade element methods, *Trans. A.S.M.E.* Ser. A, **85,** 1, 1–8 (1963).

SHEETS, H. E., The slotted blade axial flow blower, *Trans. A.S.M.E.* **78,** 8, 1683–1690 (1956).

VAN NIEKERK, C. G., Ducted fan design theory, *J. Appl. Mech.* **25,** 325–31 (1958).

VERMES, G., Influence of the degree of reaction on the efficiency of an axial compressor stage (in English), *Acta. Tech. Acad. Sci. Hung.* **17,** 25–38 (1957).

YOUNG, R. H., Contra-rotating axial flow fans, *Journal I.H.V.E.* **18,** 187, 448–77 (1951).

CENTRIFUGAL FANS

ABE, S., *Theory of the Impeller of Multi-blade Centrifugal Fans*, Rep. Inst. High Speed Mech. (Japan), **10,** 149–76 (1959).

ACOSTA, A. J. and BOWERMAN, R. D., An experimental study of centrifugal pump impellers, *Trans. A.S.M.E.* **79,** 8, 1821–39 (1957).

ANDERSON, H. H., Centrifugal pumps—an alternate theory, *Proc. I. Mech. E.* **157,** 57–84 (1947).

BLAHÓ, M. and PRESZLER, L., Investigation of the volumetric loss of a centrifugal fan (in English), *Acta. Tech. Acad. Sci. Hung.* **18,** 255–61 (1957).

BOMMES, L., The effect of the number of blades on the characteristics of a backward curved centrifugal fan (in German), *Heiz.-Luft.-Haustech.* **14,** 5, 159–166; 6, 206–9; 7, 228–33 (1963).

BOUWMAN, H. B., The influence of inlet form and bends on performance of centrifugal fans (in German), *Heiz.-Luft.-Haustech.* **11,** 7, 170–2 (1960).

BOWERMAN, R. D. and ACOSTA, A. J., Effect of the volute on the performance of a centrifugal pump impeller, *Trans. A.S.M.E.* **79,** 5, 1057–69 (1957).

FUJIE, K., Three-dimensional flow in centrifugal impellers, *Bull. J.S.M.E.* **1,** 1, 42–59; 3, 275–82 (1958).

GRUBER, J. and SZENTMARTONY, T., Study of the entry of fluid in the moving passages in centrifugal hydraulic machines (in French), *Acta. Tech. Acad. Sci. Hung.* **3,** 381–8 (1952).

GRUBER, J. and BLAHÓ, M., On the observation of streamlines in radial flow impellers (in English), *Acta. Tech. Acad. Sci. Hung.* **7**, 29–42 (1953).

GRUBER, J. and BLAHÓ, M., Measurement of pressure distribution on blades of centrifugal flow impellers (in English), *Acta. Tech. Acad. Sci. Hung.* **9**, 37–48 (1954).

GRUBER, J., Modern design recommendations for the construction of centrifugal fans (in German), *Heiz.-Luft.-Haustech.* **10**, 162–6 (1959).

HÖNMANN, W., On the problem of optimum impeller width of radial flow fans (in German), *Heiz.-Luft.-Haustech.* **12**, 161–7, 211–6 (1961).

HÖNMANN, W., Investigations into flow separation in the suction space of radial fans (in German), *Heiz.-Luft.-Haustech.* **13**, 113–9, 155–7, 176–81, 260–6 (1962).

LAAKSO, H., Improved design of fans with drum type rotors, *Engineers' Digest*, **20**, 1, 71–3 (1959).

LOWN, H. and WIESNER, F. J., Prediction of choking flow in centrifugal impellers, *Trans. A.S.M.E.* Ser. D, **81**, 29–36 (1959).

MOSER, K., Investigations into centrifugal fan spiral casings (in German), *Heiz.-Luft.-Haustech.* **8**, 12, 319–21 (1957).

MURATA, S., Research on flow in a centrifugal pump impeller, *Bull. J.S.M.E.* **5**, 17, 88–116; 18, 259–76 (1962).

PANTELL, K., Friction of discs rotating in fluid (in German), *Forsch. Ing-Wes*, **16**, 4, 97–108 (1949–50).

PANTELL, K., Blade design for turbomachinery (in German), *Konstruction*, **1**, 3, 77–82 (1949).

SHIRAKURA, M., An approximate method of determining the slip factor of the radial outward flow impeller, *Bull. J.S.M.E.* **1**, 2, 171–8 (1958).

STEPANOFF, A. J., Inlet guide vane performance of centrifugal blowers, *Trans. A.S.M.E.* Ser. A, **83**, 4, 371–80 (1961).

STEPANOFF, A. J. and STAHL, H. A., Dissimilarity laws in centrifugal pumps and blowers, *Trans. A.S.M.E.* Ser. A, **83**, 4, 381–91 (1961).

WORSTER, R. C., The flow in volutes and its effect on centrifugal pump performance, *Proc. I. Mech. E.* **177**, 31, 843–76 (1963).

MISCELLANEOUS FANS

COESTER, R., Theory and experimental investigations of the cross flow fan, *Mitt. Inst. für Aerodynamic, Zurich*, No. 28 (1959).

COESTER, R., Note on the theory of the cross flow fan (in German), *Heiz.-Luft.-Haustech.* **12**, 169 (1961).

DOWELL, M. F., A study of air movement through axial flow free air impellers, *G.E. Rev.* **42**, 5, 210–7 (1939).

FORMAN, G. W. and KELLY, N. W., Cooling tower fan performance, *Trans. A.S.M.E.* Ser. A, **83**, 2, 155–60 (1961).

DE FRIES, J. R., 66 years of cross flow fans (in German), *V.D.I. Berichte*, **38** (1959). (Vorträge de V.D.I. Tagung, Stuttgart, 1958.)

LAAKSO, H., Cross flow fans with $\psi > 4$ (in German), *Heiz.-Luft.-Haustech.* **8**, 324–5 (1957).

MURATA, S., Research on the flow in a centrifugal pump impeller (mixed flow impeller), *Bull. J.S.M.E.* **5**, 17, 110–7; 20, 683–8 (1962).

MYLES, D. J., *A Design Method for Mixed-flow Fans and Pumps*, N.E.L. Report No. 177 (1965).

REINDERS, H., The cross flow fan—a problem in air technology (in German), *Heiz.-Luft.-Haustech.* **7**, 88 (1956).

SMITH, V. J., *Air Circulator Fans: A Design Method and Experimental Studies*, Aero. Res. Lab., Melbourne, Australia, A 119 (1960).

WARD, L. G., The performance of ceiling and desk fan blades, *Electrical Energy*, pp. 212–6, March 1957.

NOISE IN FANS

BARRETT, A. J. and OSBORNE, W. C., Noise measurement in cylindrical fan ducts, *Journal I.H.V.E.* **28**, 306–18 (1960).

BEIER, K. A. and WEIR, T. J., Vehicular fan noise, *Noise Control*, **4**, 27–31 (1958).

BERANEK, L. L., REYNOLDS, J. L. and WILSON, K. E., Apparatus and procedures for predicting ventilation system noise, *J.A.S.A.* **25**, 2, 313–21 (1953).

BERANEK, L. L., KAMPERMAN, G. W. and ALLEN, C. H., Noise of centrifugal fans, *J.A.S.A.* **27**, 2, 217–9 (1955).

BOMMES, L., Noise development in fans of low and average peripheral speed (in German), *Lärmebekampfung*, **5**, 5, 69–75 (1961).

BREWER, G. A., Vibration analysis of a 700 hp induced draught fan, *Proc. Soc. Exp. Stress Anal.* **16**, 1, 17–26 (1958).

CHAN, D. O. P. and OSBORNE, W. C., Noise measurement in rectangular fan ducts, *Journal I.H.V.E.* **29**, 252–62 (1961).

DALY, B. B., Noise level in fans, *Journal I.H.V.E.* **26**, 29–44 (1958).

DALY, B. B., Fan sound control, *G.E.C. Journal*, **29**, 3, 141–151 (1962).

GUTIN, L., *On the Sound Field of a Rotating Impeller*, N.A.C.A. Tech. Rep. 1195.

HEUBNER, G. H., Noise of centrifugal fans and rotating cylinders, *A.S.H.R.A.E. Journal*, **5**, 11, 87–94 (1963).

HOWES, F. S. and REAL, R. R., Noise origin, power, and spectra of ducted centrifugal fans, *J.A.S.A.* **30**, 714–20 (1958).

HÜBNER, G., Noise in radial fans (in German), *Kampf dem Lärm.* **7**, 2, 42 (1960).

JAUMOTTE, A. L., Fan noise (in French), *Chaleur et Climats*, **27**, 320, 33–43 (1962).

JOHNSON, D. R., *Fan Noise Measurement Procedures*, H.V.R.A. Lab. Rep. 14 (1963).

KERKA, W. F., Evaluation of four methods for determining sound power output of a fan, *A.S.H.R.A.E. Journal*, **63**, 367–88 (1957).

MADISON, R. D. and GRAHAM, J. B., Noise variation with changing fan operation, *Heat. Pip. Air Cond.* **30**, 1, 207–14 (1958).

MALING, G. C. and GOLDMAN, R. B., Noise from small centrifugal fans, *Noise Control*, **1**, 6, 26–29 (1955).

MALING, G. C., Dimensional analysis of blower noise, *J.A.S.A.* **35**, 1556–64 (1963).

PIESTRUP, C. F. and WESLER, J. E., Noise of ventilating fans, *J.A.S.A.* **25**, 2, 322–326 (1953).

RIOLLET, G., *Noise Generation by a Fan by Dimensional Analysis* (in French), 9th Congrès Inter. Mecan. Appl. Univ. Bruxelles, **2**, 448–58 (1957).

ZELLER, W. and STANGE, H., Prediction of noise power from axial fans (in German), *Heiz.-Luft.-Haustech.* **8**, 322–3 (1957).

PRESSURE LOSS IN DUCTS AND FITTINGS

GILMAN, S. F., Pressure loss of divided flow fittings, *A.S.H.A.E. Trans.* **61**, 281–296 (1955).

KRATZ, A. P. and FELLOWS, J. R., *Pressure Losses Resulting from Changes on Cross-sectional Area in Air Ducts*, Univ. Ill. Eng. Exp. Station Bull. No. 300 (1938).

LOCKLIN, D. W., Energy losses in 90 degree duct elbows, *A.S.H.V.E. Trans.* **56,** 479–502 (1950).

McELROY, G. E., *Pressure Losses due to Bends and Area Changes in Mine Airways*, U.S. Bureau of Mines, I.C. 6663 (1952).

MOODY, L. F., Friction factors for pipe flow, *Trans. A.S.M.E.* **66,** 671–84 (1944).

WARD SMITH, A. J., The flow and pressure losses in smooth pipe bends of constant cross section, *J. Roy. Ae. Soc.* **67,** 437–47 (1963)

INDEX